Computer-aided drug design.
The HCV family example.

Dr Dimitrios P. Vlachakis

B.Sc., M.Sc., M.Phil., Ph.D.

ISBN Number: 978-0-557-77243-8

University of Cardiff, Wales, UK

TABLE of CONTENTS

INDEX   **PAGE NUMBER**

Table of Contents   3

Abstract   6

Abbreviations   9

Chapter 1

    Introduction   10

        1.1.1   Flaviviridae   11
        1.1.2   Taxonomy   12
        1.1.3   Genome   12
        1.1.4   Replication   14
        1.1.5   Transmission   15
        1.1.6   Viral Hepatitis C Infection and World   16
                 Prevalence
        1.1.7   The Hepatitis C virus   20
        1.1.8   The West Nile virus   26
        1.1.9   The Dengue virus   27
        1.1.10   The Yellow Fever virus   29
        1.2   The Hepatitis C virus Helicase Protein   32
        1.3   Proposed function of the Helicase   39
        1.4   Known compounds that work on the   48
                 HepC Helicase

Chapter 2

    Molecular Modelling   58

        2.1   Introduction   59
        2.2   The Hepatitis C Helicase Protein   60

Chapter 3

    Homology Modelling   69

3.1	Introduction	70
3.2	Homology Modelling of the Helicase Protein Model Series of the Flavi family	72
3.3	Homology Modelling of the Polymerase Protein Model Series of the Flavi family	76
3.4	Model Evaluation & Optimization	80
3.5	Conclusions	89

Chapter 4

Drug Design		92
4.1	Introduction	93
4.2	Drug Design for the HepC Helicase – RNA binding motif	98
4.3	Drug Design for the HepC Helicase – Application of the "tube"	109
4.4	Drug Design for the Dengue Helicase	112

Chapter 5

Molecular Docking & Screening		117
5.1	Introduction	118
5.2	Hepatitis C	120
5.3	Dengue	124

Chapter 6

Molecular Biology		128
6.1	Introduction	129
6.2	Polymerase Chain Reaction towards Gene Amplification	131
6.3	Vector Preparation I – Plasmid Restriction & Phosphatase Reaction	132
6.4	Vector Preparation II – Gene Insertion	133
6.5	Transformation	134
6.6	Microbiology	136
6.7	DNA Extraction and Cell Lysis (mini-prep)	138
6.8	DNA Purification and Concentration	140

6.9	Cloning	141
6.10	Primer Design I	145
6.11	Sequencing	146
6.12	Site Directed Mutagenesis and Primer Design II	147
6.13	Conclusion	148
6.14	Gene Sequences & Data for the Dengue Proteins	150

Chapter 7

Biochemistry		152
7.1	Introduction	153
7.2	the pET expression system	155
7.3	Protein Expression	158
7.4	Conclusions	165
References		166
Appendices		182

ABSTRACT

Hepatitis C and Dengue viruses belong to the family of Flaviviridae[1]. Viruses in this family are enveloped, have positive-sense RNA and are responsible for a variety of life threatening diseases. Hepatitis C virus is the major etiological agent of post-transfusion hepatitis worldwide[1]. An estimated 3 % of the world's population is infected with HCV according to the World Health Organization. Infection with HCV will most regularly result in chronic hepatitis, which leads to liver cirrhosis, hepatocellular carcinoma and liver failure. Dengue is currently the most important viral disease transmitted by mosquitoes afflicting humans the world context[2]. Clinical symptoms range from mild fevers to a severe haemorrhagic disease. To date neither specific antiviral treatments exist nor are there any vaccines available for both infections[3]. Thus there is an urgent need for new therapies.

The aim of this project is to design compounds that will inhibit the function of the helicases of the hepatitis C virus, the dengue virus, the Japanese Encephalitis Virus and the Yellow Fever virus (all members of the Flavi virus family). These helicases are interesting targets for drug design, firstly for their vital function in the viral cell cycle and secondly for the fact that human cells lack helicases capable of unwinding positive sense double stranded RNA. The 3D structure of the Hepatitis C helicase has been resolved and was obtained from the Protein

DataBank (1A1V and 8OHM). The structures of the Dengue, Japanese Encephalitis and the Yellow Fever helicases are unavailable and were obtained by homology modelling, using the helicase of Hepatitis C as template. A set of small libraries of compounds was designed using *de novo* drug design. The evaluation of these compounds is initially performed *in silico* and is completed *in vitro*, as soon as the best *in silico* scoring compounds are synthesized and tested in the helicase biological enzymatic assay.

The viability of the homology models of the Dengue, Japanese Encephalitis and the Yellow Fever helicases was evaluated by an *in silico* scoring function and from the fact that the RNA binding motif and

Figure 1. The model of the Helicase of Dengue-type II virus (white) superimposed on the HepC template (black)-(ssRNA is in spacefill).

the ATP binding motifs were conserved between the model and the template helicase from HepC (figure 1). Examination of the structure of the hepatitis C helicase showed that two residues (ARG393 and CYS431) are probably involved in the interaction with the oligonucleotide and are suitable to be used as

targets for drug discovery. *De novo* drug design was performed on the helicase as a whole, but with greater focus on the previous two amino acids. The structure based drug design experiment produced a series of potential lead compounds, which were evaluated by docking/scoring methods *in silico*, they were screened on the two helicases and the more promising ones were selected to be synthesized. The synthesis of these compounds is currently under development.

The establishment of the helicase assay came out using molecular biology techniques for recombinant protein expression. The genes of the Helicase and the NS3 domain (Helicase and Protease) were incorporated into the high copy pGEM-T plasmid in order to amplify them. Two sets of DH5-alpha strains of E.Coli bacteria cells were transformed with the designed vectors and cultured in liquid suspensions to generate stocks of clones. The vectors were extracted from the DH5-alpha strains and the genes were digested and ligated into the pET23b+ expression vector into the BL21lys E.Coli strain. The genes were induced and the protein was produced as revealed by SDS-PAGE. Due to a misplaced stop-codon at the end of the two genes the His-tag of the pET vector was pushed out of frame and protein isolation was impossible. Currently work is ongoing in an attempt to remove the stop-codon, place the His-tag in frame and isolate the proteins by column chromatography.

Keywords: molecular modelling, de novo drug-design, homology modelling, model evaluation, docking, chemical synthesis, cloning, recombinant protein, biological assay

ABBREVIATIONS

LPC	:	Ligand Protein Contacts
HCV	:	Hepatitis C Virus
SAH	:	S-Adenosylhomocysteine Hydrolase
SCR	:	Structurally Conserved Regions
SVR	:	Structurally Variable Regions
AS	:	Active Site
RMSd	:	Root Mean Square deviation
MOE	:	Molecular Operating Environment
PDB	:	Protein Data Bank
HTS	:	High Throughput Screening
HepA	:	Hepatitis A
HepB	:	Hepatitis B
HepC	:	Hepatitis C
RP	:	Ramachandran Plot
CSU	:	Contacts of Structural Units
WNV	:	West Nile Virus
JE	:	Japanese Encephalitis
YF	:	Yellow Fever
NCI	:	National Cancer Institute
PPS	:	Polyphosphoric Acid
PCR	:	Polymerase Chain Reaction

CHAPTER 1:

INTRODUCTION

1.1.1   Flaviviridae:

Flaviviridae is a family of viruses that infect vertebrates. Distinct viral structures of this family are visible in thin sections of infected tissue. The size of virion has been estimated by filtration. Virions of the flaviviridae family are enveloped and slightly pleomorphic during their life cycle. They are spherical in shape and usually 40-60 nm in diameter. Their nucleocapsids are isometric and sometimes penetrated by stain. The usual size of the nucleocapsids is 25-30 nm in diameter and they have polyhedral symmetry. Virions of the flaviviridae family contain one molecule of linear positive-sense single stranded RNA. The total genome length is 9500-12500 nt. The 5' end of the genome has a cap, or a genome-linked protein (VPg). The 3' end regularly has no poly (A) tract (except some strains of tick-borne encephalitis complex of flaviviruses, which have a poly (A) tract). Their nucleic acid material is fully Encapsidated and solely genomic. The genome of flaviviridae features a 5' end that encodes structural proteins, whereas the non-structural proteins including protease, helicase and polymerase, are encode at the 3' end.

1.1.2   Taxonomy:

Flaviviruses have been subdivided by the ICTV into three genera:

- **Genus:** *Flavivirus*
- **Genus:** *Pestivirus*
- **Genus:** *Hepacivirus*

The *Flavivirus* family contains several nasty members including yellow fever virus, dengue fever virus, and Japanese encaphilitis (JE) virus. The *Pestivirus* genus consists mainly of the three serotypes of bovine viral diarrhea, but no known human pathogens. Finally, the genus of *Hepacivirus* is home to the hepatitis C virus and its sub-family.

1.1.3   Genome:

Genus Flavivirus genomes consist of a single piece of linear, single-stranded but positive sense RNA. Just because the viral RNA has positive sense, the nucleic acid itself is capable of causing an infection in the appropriate host cells. The total genome can range from 10 to 12 kilobase pairs. The 3' terminus of the viral genome is not polyadenylated, whereas the 5' end has got a methylated nucleotide cap, which apparently allows the translation process to occur. Sometimes it is possible to get a genome-linked protein (VPg) in the place of the methylated nucleotide cap.

The genome of the *Pestivirus* family is reported to be approximately 12.5 kb in length. The same stands for the *Flavivirus* genus too. Moreover, no poly-A tail exists on the 3' end of the RNA, however, *Pestivirus* genus members *lack* a 5' cap. In both genera, usually structural genes are located towards the 5' end of the RNA.

The *Pestivirus and* the *Hepacivirus* genus have internal ribosomal entry sites (IRES), which are responsible for providing a site for the initiation of the translation process for host ribosomes. On the other hand the *Flavivirus* genus does not have internal ribosomal entry sites, but is capable of scanning the ribosomes to begin protein synthesis.

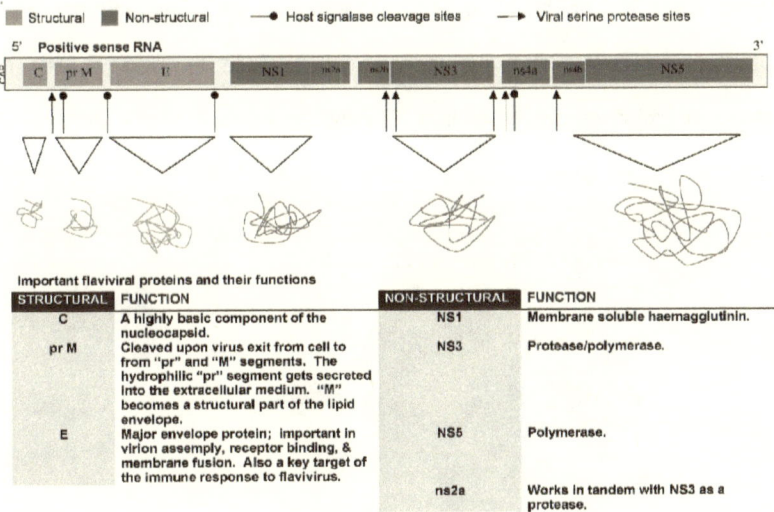

Figure 2. As shown above, the flavivirus genome consists of a single strand of positive sense RNA. Upon translation several enzymes cut the single polypeptide into functional protein units. Some of the proteases are actually provided by the host cell.

1.1.4 Replication

The replication process takes place in the cytoplasm and lasts approximately 20-30 hours for flaviviruses. The Cellular receptors are still not well known, but there is evidence that they must be distributed in the body to account for the various sites of simultaneous flavivirous infection. The viral genome is firstly translated as a single polyprotein, which will later on gets cleaved into individual-mature proteins. The complementary negative strand of the RNA is synthesized by NS proteins. This strand is used as a template structure for the genomic progeny RNA synthesis process. The viral assembly takes place during budding, into cytoplasmic vacuoles and not as it was until recently thought in the cell surface as occurs with its arbovirus relative, the togavirus. Finally the release of the Virions will occur when cell lyses.

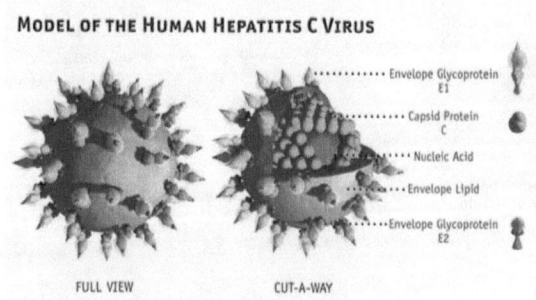

Figure 3. The Hepatitis C Virus

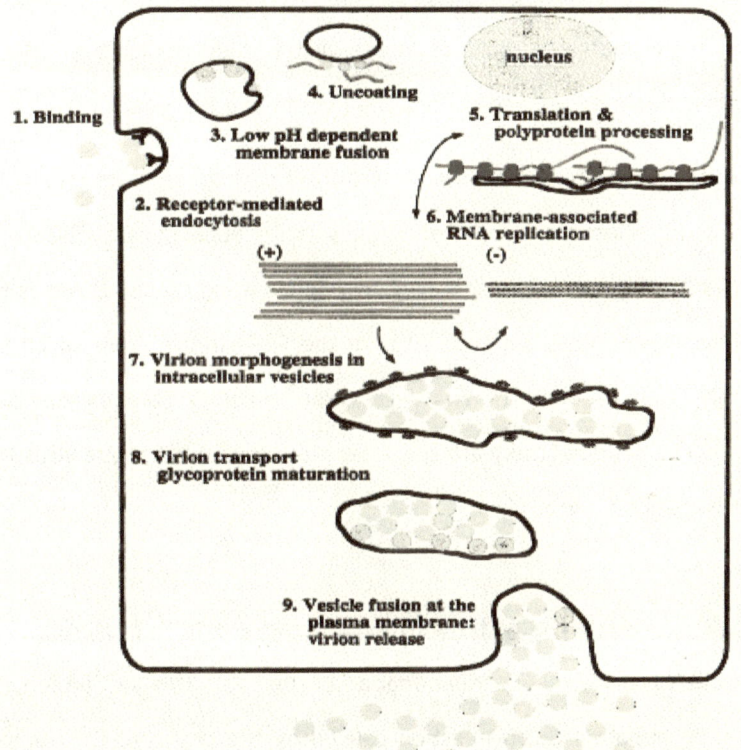

Figure 4. The generic life cycle of the viruses in the Flaviviridae family, upon infection of host cell (reference: www.google.com).

1.1.5   Transmission

All flaviviruses are carried and transmitted by arthropods (commonly mosquitoes and ticks), while the hepatitis C virus is spread parenterally. Most commonly via contaminated bodily fluids. A vital characteristic feature for viral transmission in the flaviviridae family is that these viruses are capable of reproducing in their own vector. If they were unable to replicate in the vector, they wouldn't be able to remain viable to be passed from one host to another one.

The flaviviridae family counts a total of 69 pathogens in its ranks. So, Flaviviridae include a cast of viruses that cause severe disease in humans. The most representative example among these viruses is the Yellow Fever Virus, the type virus of the Flaviviridae, from which the family takes its name (*flavus* = Latin for "yellow"). Famous members of the Flaviviridae family are the West Nile virus, the Dengue virus and the Hepatitis C virus.

1.1.6   Viral Hepatitis C Infection and World Prevalence

Hepatitis literally means inflammation of the liver. This section focuses on viral hepatitis C, an infection caused by the hepatitis C virus that primarily affects the liver. The hepatitis C virus is estimated to have infected 170 million people worldwide[13], with most countries reporting a prevalence of 1.0% to 4.9%. Currently, about 40% to 60% of chronic liver disease is because of HCV. Between 50 – 80 percent of patients infected with HCV are going to develop chronic infection. 20% of them will develop liver cirrhosis within 20 years [41]. HCV infection is a considered to be rather rare in terms of the overall number of cases

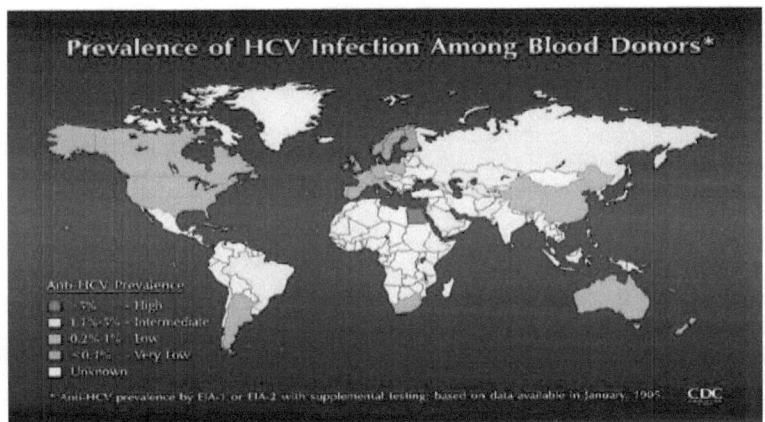

Figure 5. World Prevalence of the Hepatitis C infection. It estimated that 3% of the world is infected with the hepatitis C virus alone today, according to the World Health Organization [36].

viral hepatitis in developed countries. Hepatitis C is currently the number one cause of chronic hepatitis in the united states of America. It is estimated that it causes around 10,000 deaths per year. Historical screens reveal that the prevalence of HCV seropositivity in the USA has increased from approximately 1% in the 1940s to approximately 1.7% nowadays. It is believed that this is due to blood transfusion, transplants and increased drug abuse.

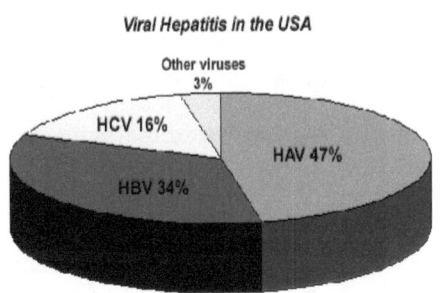

Figure 6. Percentages of the HAV, HBV and HCV infections in the USA

(reference: www.google.com).

17

The most prevalent route of infection seems to be the parenteral, the higher rates of infection have been observed in high-risk groups such as intravenous drug abusers, hemophiliacs and recipients of unscreened blood transfusions. The probability of sexual transmission exists but still risk factor appears to be very low. Finally the transmission of HCV via the vertical route (mother → neborne) takes place at a rate of 5-10%. Previous knowledge of the disease lowers the risks by operating a caesarean section to the mother.

After infection with HCV almost all patients will develop a vigorous antibody and cell-mediated immune response that will fail to cope with the infection, but may contribute towards liver damage. It has been found that the majority of the flavivirus infections are cytopathic. Unfortunately in the case of HCV this is not clear, because the virus cannot be cultured. The spontaneous resolution of the chronic liver disease does not happen very often (<2%). All patients with chronic Hepatitis C are at risk of developing hepatocellular carcinoma (HCC). Studies have revealed that sometimes and in specific populations the infection with HCV may have a more benign outcome.

Time it takes from HCV after transmission to develop cancer varies between 10 to 50 years (most usually about 30 years). The association between chronic HCV infection, cirrhosis, and hepatocarcinogenesis is very strong (Haydon et al Gut 40: 128-132, 1997; Degos et al Gut 47:

131-136, 2000). The duration from the start of the acute hepatitis until the time of liver cirrhosis diagnosis and the time of HCC diagnosis is approximately 20 and 30 years, respectively. The acute phase of the disease will lasts for 2 to 3 years after transmission. After that and for about 10 to 15 years the "silent phase" of the disease will follow.

It is presently unclear how an RNA virus, such as HCV, establishes a persistent infection, because any little is known about the biology of this virus. The virus has a very rapid turnover of plasma virus content in patients. It is estimated that the half-lives of the particles ranges between 100 to 182 min.

The core of the HCV protein interacts and represses transcription from the p53 promoter. As a result it will block the synthesis of p53. It has not been confirmed yet but it is believed that this may be important for the survival of the hepatocarcinoma cells that have been transformed by the Hepatitis C virus. Transgenic mice that have been made to express the core protein of HCV will soon develop HCC. This is mainly because of the formation of intracellular reactive oxygen species, which result soon after mitochondrial malfunction and injury.

The discovery of HCV was done using molecular biology methods. This lead even to the sequencing of the entire genome, but still a permissive cell culture system for HCV has not been established. This is a big disadvantage in the fight with the disease since the production of specific drugs against HCV cannot be fully investigated, because no efficient diagnostic method has been developed.

1.1.7 The Hepatitis C Virus

The genome of HCV was first cloned in 1989. The genome of the HCV is 9.4 kb long, contains a single long open reading frame which is responsible for encoding a polyprotein of 3,010 amino acids [15]. There is a non-coding region (NCR) of 324-341 nucleotides at the 5' end & on the 3' end there is a NCR of variable length including a poly (U) tract. The 5' NCR contains an IRES that is very similar in function (not structure) to that of picornaviruses. The nucleotidic sequence of the Hepatitis C virus is highly variable; the most distant strains have only 60% nucleotide sequence homology. Strains from around the world have now been classified into 6 main categories, each with several subtypes, based on sequencing properties (Simmonds et al, Hepatology 21: 570-83, 1995). Categories 1, 2 and3 account for almost all infections in Europe. Category 4 is prevalent in Egypt & Zaire, category 5 is prevalent in South Africa finally category 6 is found in Hong Kong. It is not clear yet if immunity to one category automatically defends infection with another strain from another category. There is though evidence that various genome types are pretty much different in their biological properties.

The non-structural region of the NS3 domain of the HCV ranges between 1027 and 1657 of the polyprotein [38]. The genome of the HCV is a positive-stranded RNA virus [13]. The 3010 amino acid long polyprotein of the virus is processed by cellular and virus-encoded proteases [16].

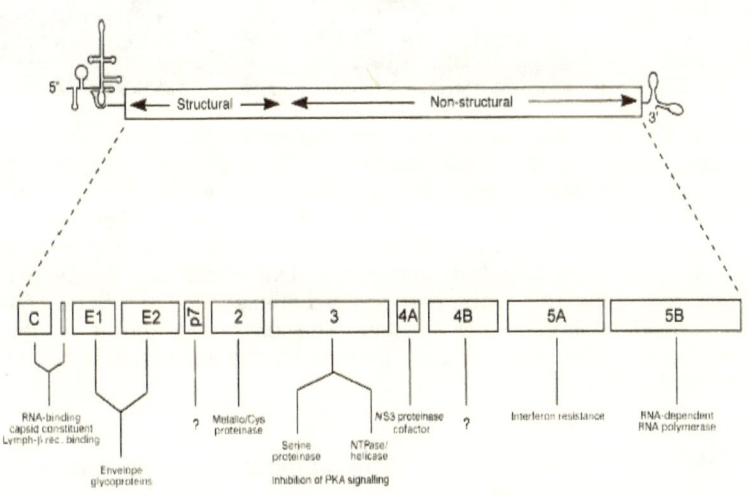

Figure 7. Hepatitis C genome organization, polyprotein processing and protein properties. On the top there is a schematic representation of the HCV genome, below are the polyprotein cleavage products and defined functions as described underneath (reference: © 1999 Blackwell Science ltd. *Journal of Viral Hepatitis*, 6, 165-181).

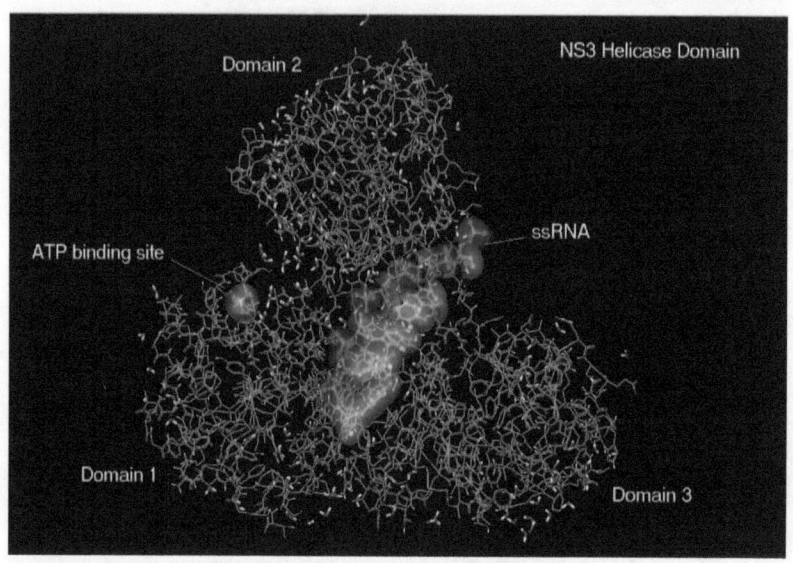

Figure 8. The upper section of the figure below (figure 2) is the crystal structure of the hepatitis NS3 helicase domain. The oligonucleotide binds with no sequence-specific interactions in a channel between the domain 1-2 and 3 of the HepC helicase. The single strand of nucleic acids is lying in a channel that is 16 Å wide and separates domains 1&2 from domain 3. The interaction between the nucleic acids and the protein is mainly between the backbone of the nucleic acid chain and the protein. The main contributor to these interactions is the symmetrically located equivalent residues between domains 1 and 2.

The life cycle of the Hepatitis C virus is typical of all RNA viruses. The first consideration of the hepatitis C virus is to attach itself and infect liver cells. Only then the virus is capable of initiating its life cycle and reproducing. Very little is known about how the natural processes of hepatitis C virus infection develops, but with the few data available the life cycle of HepC can be summarized in 5 different steps:

STEP 1: First thing that the virus has to do is to find and attach itself to a liver cell. Then the Hepatitis C virus will utilize special proteins present on its protective lipid coat in order to help it attach to a receptor site on the host liver cell.

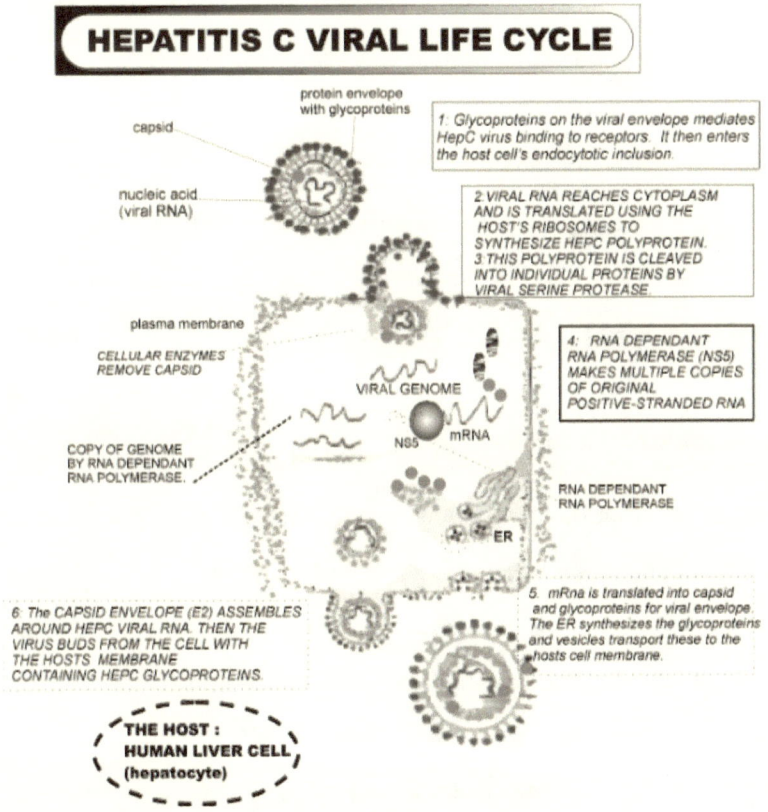

Figure 9. The life cycle of Hepatitis C.

STEP 2: The viral protein core will penetrate host cell's plasma membrane and will enter into the cytoplasm of the liver cell. In order to penetrate the host cell membrane HCV will make use of its protective lipid (fatty) coat. HCV will attempt to merge its lipid coat with the host's outer membrane. As soon as the lipid coat has successfully fused to the plasma membrane, the membrane of the host will engulfs the virus (a process identical to endocytosis). When the virus is inside the host its protein coat dissolves. As a result the viral RNA is now released into the cell. It is not clear yet if the virus first enters the host and then dissociation of its protein coat occurs or if during penetration of the cell membrane the protein coat is broken open (after fusion with the liver host cell) and then the contents of the virus are released into the cytoplasm. There is the option that there may be enzymes on the liver cells cell membrane that the HCV may utilize to dissolve its protein coat.

STEP 3: Now that the viral RNA is in the cytoplasm it will find the cell's ribosomes and it will begin the process of the production of materials necessary for the viral reproduction. Hepatitis C has got a "sense" strand of RNA. So the viral RNA itself can straightaway be read by the host cell's ribosomes – as if it was normal mRNA. During this process two things take place simultaneously. Firstly the virus begins to produce the materials coded in its RNA, but it also influences most of the normal functions of the cell, in an attempt to drive the cell crazy. The aim of this taking over behaviour is that the virus wants to stop the liver cell from using its resources so that they are all available for the needs of the virus. Another interesting feature of HCV is that it will also try to push the host cell to reproduce, in an attempt to create a new host for viral reproduction. This may be why HCV infection is directly associated with hepatocellular carcinoma. The first product of the viral RNA is the RNA transcriptase that it will use for reproduction.

STEP 4: As soon as the RNA transcriptase is made, it will make an antisense version of the viral RNA, to be used later on as a template for the generation of new viral RNA. Viral RNA will be copied thousands of times in order to make genetic material for new viruses. It is certain that this new RNA will contain various point mutations. This is a powerful weapon of the HCV to escape from the host's immune system response. The Viral RNA will drive the production of protein-based capsomeres to be used for the new viruses' protein coats. These proteins are made by ribosomes. These proteins will assemble around the new viral RNA. The final product is known as nucleocapsid.

STEP 5: The new viruses will go to the inside of the plasma membrane and will interact with it. This interaction will help them to attach to it and finally to establish a bud. The plasma membrane will surround the virus and it will release it on the outside of the cell. During the release the virus will take with it a lipid coat from the host's cell membrane. This will help the virus to attach to another liver cell further down. This process will continue until the host cell will die from exhaustion.

1.1.8   The West Nile Virus

West Nile fever is caused by a virus that is part of the Flaviviridae family. There are nearly 70 different viruses in this group, formerly termed group B arboviruses, of which nearly half are known to cause illness in humans. The World Health Organization defines arboviruses (arthropod-borne viruses) as a group as those "which are maintained in nature principally, or to an important extent, through biological transmission between susceptible vertebrate hosts by hematophagous arthropods; they multiply and produce viremia in the vertebrates, multiply in the tissues of arthropods, and are passed on to new verte-brates by the bites of arthropods after a period of extrinsic incubation" (Sanford, 1991). Common viruses in this classification, in addition to West Nile, include yellow fever, dengue, Japanese encephalitis, St. Louis encephalitis, and tick-borne encephalitis viruses. These viruses are generally spread by mosquitoes or ticks; human-to-human spread does not occur. Infection with these viruses does not produce a unique clinical picture. Therefore, travel to an endemic area and laboratory tests are important for identifying a specific infection.

West Nile virus is a mosquito-borne virus found most commonly in Africa, France, India, Indonesia, the Middle East, and Soviet countries. In 1999, a West-Nile-like virus was identified in patients living in the Northeast United States. The bird is the primary host and the principal vector is *Culex univittatus*. However, other mosquitoes are known to carry the virus, including *Culex pipiens, Culex antennatus*, and *Culex tritaeniorhynchus* (Asia). Other animal reservoirs are not part of the virus's normal life cycle.

West Nile fever is common in the Middle East with most individuals exposed as children. Children experience a nondescript viral illness with fever that is rarely diagnosed. Neighboring Israel also experiences infection although there, it is more likely the young adult than the child who becomes infected. Spread occurs primarily in the summer months when the mosquito population increases.

1.1.9   The Dengue Virus

Dengue fever is caused by a virus that is part of the Flaviviridae family. There are nearly 70 different viruses in this group, formerly termed group B arboviruses, of which nearly half are known to cause illness in humans. Other common viruses in this classification include yellow fever, West Nile, Japanese encephalitis, St. Louis encephalitis, and tick-borne encephalitis viruses. The most common infection in humans is caused by the dengue virus, of which there are four types. Flaviviruses are generally spread by mosquitoes or ticks; human-to-human spread does not occur. Infection with these viruses does not produce a unique clinical picture. Therefore, travel to an endemic area and laboratory tests are important for identifying specific infection.

Dengue and dengue hemorrhagic fever (DHF) are caused by infection with one of four antigenically distinct, virus serotypes (DEN-1, DEN-2, DEN-3, and DEN4). Once infected with one of these serotypes, the individual develops specific immunity. However, cross-immunity does not develop. It is theoretically possible, therefore, for an individual to be infected four times, each time with a different serotype.

Dengue is mostly seen in tropical urban areas. As with other members of the Flaviviridae family, the virus is transmitted through mosquito bites, specifically *Aedes aegypti*. This mosquito, a domestic, day-biting mosquito, prefers to feed on humans (Gubler and Clark, 1995). In some parts of the world (mostly Asia and Oceania) other vectors have been implicated: *A. albopictus, A. scutellaris, and A. polynesiensis*. Dengue is the most important mosquito-borne viral disease, affecting humans with a distribution comparable to that of malaria. Approximately 2.5 billion people are living in areas at risk for epidemic transmission (Gubler and Clark, 1995). Tens of millions of cases of dengue fever occur annually along with up to hundreds of thousands of cases of dengue hemorrhagic fever.

Dengue hemorrhagic fever is the most serious manifestation of the disease. This process, an immunologic reaction, occurs for the most part in individuals already sensitized to the disease, either actively through infection or passively in infants through placental transfer of immunoglobulin from mother to child. Initially, dengue hemorrhagic fever appears the same as dengue but after several days the patient deteriorates with prostration, restlessness, signs of circulatory collapse (diaphoresis, cold extremities, dyspnea, circumoral and peripheral cyanosis, and hemorrhagic manifestations). Available laboratory tests cannot identify who will ultimately develop this manifestation.

1.1.10   The Yellow Fever Virus

Yellow fever is a viral hemorrhagic fever which strikes an estimated 200 000 persons worldwide each year and causes an estimated 30 000 deaths. The case fatality rate may reach 20% to 80%; however, these figures are based on the most severe cases that are hospitalized and the overall case fatality rate is lower.

The yellow fever virus is small (35 to 45 nm) and consists of a core containing single-stranded RNA surrounded by a lipid envelope. The genome has been completely sequenced and found to contain 10 862 nucleotides *(Rice et al. 1985)*. The envelope contains a single glycoprotein with type and group-specific antigenic determinants. Yellow fever virus can be inactivated with lipid solvents (ether, chloroform), heat (56°C for 30 minutes), and ultraviolet light *(Monath 1990).*

Antigenic differences have been shown between strains of yellow fever virus. Antibody-absorption techniques can distinguish between strains from South America and Africa *(Clarke 1960).* Strains can also be differentiated on the basis of virulence characteristics for mice *(Fitzgeorge, Bradish 1980).* RNA oligonucleotide mapping has shown three genetically distinct geographical variants in Africa: Senegal-Gambia; Ivory Coast-Burkina Faso-Nigeria; Central and East Africa *(Deubel et al. 1986).*

The disease yellow fever was first distinguished from malaria, dengue, and other tropical diseases during the 1647 to 1649 epidemics in Barbados, Cuba, Guadeloupe and Mexico *(Bres 1986).* Since then, it has raged as periodic epidemics in the Americas and Africa. In 1900, a commission headed by Walter Reed confirmed that the disease was transmitted from human to human by the mosquito *Aedes aegypti,* a finding earlier theorized by the Cuban physician Carlos Finlay in 1881. This information led to efforts at mosquito control in the Americas, with excellent results in eliminating the disease from many areas.

There are two epidemiologic patterns of yellow fever virus transmission: the urban cycle and the forest cycle (also known as the jungle or sylvatic cycle). The two epidemiologic patterns lead to clinically identical disease, since they are produced by the same virus. In some instances, spread from forest to urban cycles has been documented. Today, the yellow fever virus circulates in an endemic, forest cycle in the Americas, resulting in up to 500 reports of infections of unimmunized forest workers per year. In Africa, yellow fever virus circulates in both urban and forest cycles, and the disease periodically explodes out of its endemic pattern to infect large number of unimmunized persons during major epidemics.

In the urban pattern, the virus is transmitted by mosquito from infected humans to susceptible humans. For the urban cycle, the mosquito vector is usually A. *aegypti,* a domestic mosquito that breeds near houses, with the female preferring to lay eggs in water collected in water jars, old tires, gutters, or discarded tin or plastic containers. In 1978, it was found that A. *aegypti* females could transmit yellow fever virus transovarially to a small proportion of their offspring and these eggs can thus maintain the virus during the dry season *(Aitken et al. 1979).* The virus multiplies in the mosquito vector. About 12 to 21 days after biting an infected person or monkey the mosquito becomes infective and it remains infective for the rest of its life.

The disease in humans is characterized by sudden onset of fever, headache, backache, general muscle pain, nausea, and vomiting. As the disease continues, albuminuria, oliguria (even anuria), and jaundice occur. Hemorrhagic symptoms may include epistaxis, hematemesis, and melena.

1.2   The Hepatitis C virus Helicase Protein

The Hepatitis C helicase consists of a 456 amino acid polypeptide. It has three domains in total, which are separated by two channels (Figure 8). The first and third domains are interacting much more together that they do with domain two. The outcome of this is that the channel between domains 1-2 and 2-3 is larger that the channel between domains 1-3 and 2-3. Domain two is supposed to undergo significant movements compared to the other two domains, during the unwinding of double-stranded nucleic acids. So, the position of domain two is very flexible relatively with the other two domains and the Helicase and acquire the form of a dynamic "hinge" that moves accordingly to the needs of the protein and the process it is involved into.

The topology of the first and the second domains is very similar. These two domains contain the structurally conserved regions of helicases of this family. This is confirmed by the superimposition of the two domains, which gives an RMSd of 2.0 Å for 76 Ca atoms. The third domain consists mostly of a-helices and is linked to domain two with a set of antiparallel β strands.

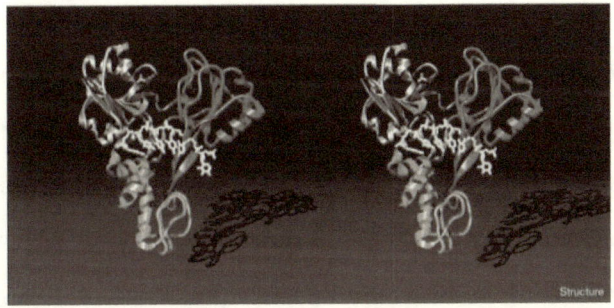

Figure 10. The NS3 Helicase domain complexed with ssDNA. Domain 1 is blue, domain 2 is red, domain 3 is green and the ssDNA is yellow. Sulphate ion is green in domain 1 (blue).

Domain 3 includes a 40 amino-acid long region, just before the a-helix in the C terminus that does not contain any 2ndary structures. This may contribute to the flexibility required by the protein during its cleavage from the NS4A domain during polyprotein processing. On the other hand towards the N-terminus of the protein there is the highly conserved "Walker A box" or "P-loop" motif. This motif is very often found among helicases and is basically a glycine rich region of the protein that provides a quite flexible loop between beta-strands and alpha-helices. The "Walker A box" has got phosphate-binding properties and is found in most ATPases. The sulfate ion (shown in figure 10) interacts with the Nitrogens Gly207 and Gly209, and the side-chains of Ser208, Lys210 and Ser211. Lys210 establishes a H_2O mediated interaction with As290 of the DExH motif (Asp-Glu-x-His). The position of the sulphate ion was found to be very similar to the position that the β-phosphate of ADP would take in the PcrA helicase–ADP complex. So, it is suggested that this the space that β-phosphate should take when NTP or nucleotide diphosphate (NDP) is bound to the HCV helicase. The residues Gln460, Arg464 and Arg467 are highly conserved residues from domain 2 that are exposed to solvent in the major channel of the Helicase. Arg461 and Arg462 are buries amino-acids in the core of the 2nd domain.

The single strand of DNA is located in the main channel of the helicase between domains 3 and 1-2 (figure 10). The size of the channel is approximately 16 Å in diameter. 5' end of the oligonucleotide is towards the part of the channel between domains 2 and 3 and the 3' end of the oligonucleotide is towards the part of the channel between domains 1 and 3.

The interactions between the ssDNA and the Helicase is basically between the backbone of the DNA, since it is nonspecific protein-DNA interaction. The majority of the established interactions are located towards the two ends of the oligonucleotide. From the protein's point of view most interactions arise from regions lacking secondary structures in domains one and two. The positioning of the interaction-participating amino acids is symmetric and as a result it appears to be a symmetric distribution of the interactions between the DNA and the protein. After superimposing the first and the second domains it was proven that the residues involved in the phosphate contacts are structurally equivalent. Furthermore the phosphate-binding amino acid series of Ser231, Thr269, Ser370 and Thr411 are conserved in NS3 domains and this is evidence that these two domains may descend from a gene duplication event. Val432 and Trp501 are also highly conserved residues among HCV NS3 sequences, nevertheless neither seems to play any role in nucleic acid binding or duplex unwinding.

The second domain has got two extended antiparallel strands (residues 430-452), which interact with the 5' of the oligonucleotide (Figure 1). This is also known as the L-45 loop and it belongs to the family of nucleic-acid binding motifs. The positioning of the domains that make the channel for the DNA/RNA in the helicase is very similar to that of the domains in the replication protein A (RPA) [32]. In both cases interactions are better formed towards the ends of the oligonucleotide and only minor interaction occur with the nucleotides in the middle. There is even an L-45 loop in RPA as well.

There are not any sequence-specific interaction with the DNA bases and the helicase. This was anticipated from the biological activity of the helicase and its behavior during enzymatic assays. Any differences in the binding affinity between different nucleotides may be due to differences in DNA distortion and base stacking.

HCV strains appear to have very high sequence conservation among them. The identity percentage of the sequence alignments is calculated to be even higher than eighty percent. Comparison of the sequences of *Bacillus stearothermophilus* [25] and *Escherichia coli* Rep DNA helicase [26] showed that there is an overall structure similarity between the domains 1A-2A of the above strains and 1-2 domains on the HepC helicase. The alignment of the amino-acid primary sequence may not be identical, but the alignment of the motifs is evidence that the arrangement of these different proteins in space and function must be similar to the each other. Site-specific mutagenesis has revealed that the function of the residues in these motifs of various helicases (including HepC) is crucial to the function of the helicases. Any change of these amino-acids will result in a protein mutant with lower affinity in unwinding dsRNA and dsDNA.

The fold of domain 1 is very similar structurally to that of common ATP transphosphorylases (for example adenylate and thymidine kinases). The GSGKT motif in domain 1 is conserved to the same loop in kinases, where its role involves the binding the β-phosphate of ATP. Site mutagenesis studies of that motif have reported that the mutant protein is inactive. Similar to the GSGKT motif is the DExH motif (motif II). The DExH motif is responsible for the binding of the Mg^{2+}–ATP substrate. Studies in adenylate and thymidine kinases revealed that an aspartate binds the Mg^{2+} and helps to establish the optimum orientation of ATP for nucleophilic attack [37,38]. Mutating this aspartate to any other amino acid will produce a helicase incapable of hydrolyzing ATP. Also His293 is essential for the hydrolysis of ATP. Mutation of His293 will lead to an inactive helicase, but capable of hydrolyzing ATP. It is suggested that this residue plays a key role between the ATP hydrolysis and nucleotide binding process.

The role of the QRxGRxGR motif is not clear yet, but this motif appears in most of the helicases of this family. Site mutagenesis studies revealed that mutation of this motif in vaccinia virus helicase and in eIF-4A will produce a helicase with less ATPase activity [41,43].

Motif VI consists of three conserved arginines [24]. All three arginines are involved in the binding of ssRNA in the helicase' s channel between domains 1 and 2. Apparently, there is distance in beliefs about the importance of these arginines as Kim suggests that they do not play such an important role in ssRNA binding. Kim suggests that Arg461 is hydrogen bonded to Asp412 and Asp412 interacts with ssRNA. So, there is an indirect importance of the Arg461, which is to keep the Asp412 in correct orientation for the RNA to be able to interact with it.

Mutagenesis studies of the arginines revealed that the resulting helicase has decreased RNA binding affinity.

Arg464 and Arg467 are expected to interact with ATP from mutation studies. Mutating the arginines to alanine or glutamine in vaccinia NPH-II or eIF-4A reduced the ATPase activity by a factor of 20% [43,44]. Arg467 is found to be concerved among all the superfamilies of helicases [22].

Motif III is located in-between the first and the second domains. Its role is to operate as a "hinge" offering the all important flexibility to the helicase protein. The Ia motif constitutes part of the beta sheet in the 1st domain, while interacting with the oligonucleotide too. Motif V is also in contact with the oligonucleotide. Thr411 hydrogen bonds to a phosphate of an oligonucleotide. The conserved motif IV (from known helicases) is absent from the HepC helicase though. There has been extensive work done on the motif IV of the superfamily I and II of helicases, but it was either done using weak criteria for the alignments or the HepC helicase evolved in a different fashion that the rest members of the two superfamilies. In the rest DNA helicases of these superfamilies of helicases motif IV is involved in the binding of ATP [25,26].

Residues in the HepC helicase that should ideally constitute part of the motif IV are the residues Ser370 and Lys371. Ser370 interacts with the oligonucleotide via a water-mediated hydrogen bond and Lys371 establishes a backbone interaction. As a result it is obvious that the HepC helicase is completely missing the motif IV of the other helicases of the I and II superfamilies of helicases.

1.3 Proposed function of the Helicase

The conserved motif VI is found across the inner part of the channel in HepC helicase, but it extends until the ATP site. The same pattern follow the structures of the adenylate kinases. In the group of adenylate kinases the conserved motif starts from the channel, where the oligonucleotide binds and extend until the ATP binding site. The purpose for this is that the binding of ATP or ATP analogues will have an immediate and direct allosteric effect on the helicase. So, the energy of the hydrolysis of ATP is utilized directly for the purposes of the enzyme, without losing any of it in exchanges between different motifs or "loose connections".

The binding of the ATP (or its analogs) will result to a conformational change in the structure of the enzyme. The most significant movements include the burial of the phosphates that were exposed to the solvent. Mutation studies on these phosphates have resulted in a wider configuration of the helicase molecule with much lower affinity for unwinding double stranded nucleotides [46 & 47].

According to KIM et al. upon ATP binding, the domains one and two get closer driven by the conformational change initiated originally in motif VI. This pattern would be expected to be common in all superfamily I and II members due to the consistency of the conserved motif VI.

The residues Gln460 and His293 belong to the motif VI and stand on opposite sites in the oligonucleotide-binding channel. These two residues may be the ones responsible for regulating the equilibrium between the "tense" and "relaxed" upon binding of the polynucleotide. The importance of these two amino acids was suspected by the fact that helicases with the DExH motif II often have a glutamine residue in motif VI, whereas helicases that have a DEAD motif often come with a histidine residue in this position [22].

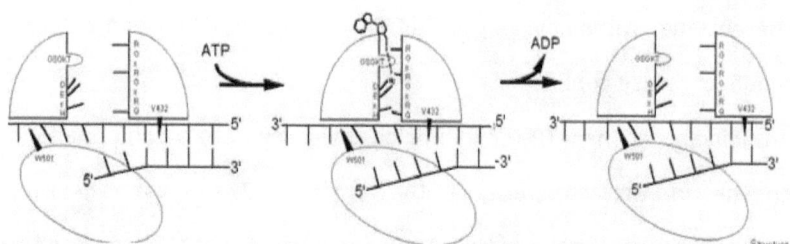

Figure 11. The schematic mechanism of RNA unwinding with the HepC Helicase. The binding of ATP will initiate the movement of domains one and two and the opening of the RNA binding channel, so that the ssRNA can translocate in the 5' to 3' direction [68].

The suspicion that the 2nd domain is flexibly linked to the rest of the protein was confirmed from the work of Yao et al [24]. The 2nd domain of the hepatitis C virus interacts directly with the polynucleotide. The movement of the above domain could influence a relative movement of the polynucleotide as well (compared to the rest of the protein). The residues Val432 and Thr448 may interact with the nucleotide bases at the 5′ end of the single-stranded nucleic acid, and this can result in a movement of the polynucleotide from the 5′ towards the 3′ direction – during the closure of the 2nd domain. Trp501 helps the single stranded nucleic acid to give up on its interaction with its surrounding residues. The positioning of Trp501 will only help the movement of the polynucleotide towards the 5′ direction. The hydrolysis of ATP and the release of ADP will result in a slight opening of the polynucleotide channel and a coordinated movement of the whole domain 2 of the helicase towards the 5′ direction.

So the hydrolysis of ATP will result in the movement of the polynucleotide relative to the rest of the helicase. Apparently, it has been found that helicases are capable of moving several bases from the polynucleotide per ATP molecule that is hydrolyzed to ADP [53].

As mentioned before the HepC helicase is very similar to other helicase (structurally), from superfamilies I and II. So, instead of just sharing sequence similarities amongst them, the HCV helicase shares a lot of conserved common motifs in its structure. There are two motifs that have been found to be extremely vital for the winding of the double stranded nucleic acids in the helicases. The first motif is the YRGXDV structurally conserved motif and the second one is the DFSLDPTF structurally conserved motif. The YRGXDV motif is the link between the IV and the V motifs in the superfamily II helicases. At the end of this loop lies the residue Arg393. Mutagenesis studies have shown that upon mutation of the residue of Arg393 to Ala, the recombinant protein is incapable of unwinding double stranded polynucleotides. Combined with the fact that this Arginine residue (Arg393) is fully exposed to the solvent and the nature of the Arginine amino acid, it makes a very good target for drug design experiments in an attempt to design compounds that will inhibit the function of the HCV helicase. The affinity of DNA winding is not completely lost to the mutant helicase, but it definitely extensively reduced.

Motif DFSLDPTF is the link of the two anti-parallel beta sheets between the V and VI motifs. Mutagenesis studies for this motif involve the manipulation and mutation of the residue Phe444. Mutating Phe444 to Ala will produce a protein that is capable of hydrolyzing ATP, but incapable of unwinding dsRNA. The affinity for unwinding dsDNA has dropped to half compared to the original helicase.

Those two motifs are known as Arg-clamp and Phe-loop respectively among all helicases of HCV (including various genotypes and quasispecies). What makes those two motifs unique, is the fact that they only appear on HepC helicases and not on the other helicases of the superfamily I and II.

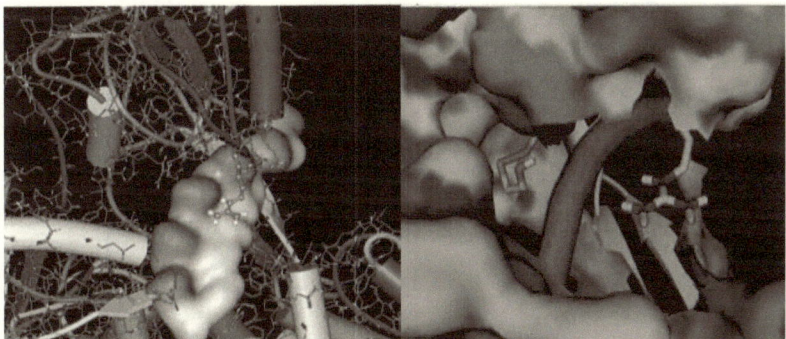

Figure 12. The HepC helicase with the ARG393 residue shown in sticks representation. It is obvious that the function of this Arginine residue in holding the oligonucleotide in place is essential.

This characteristic gives those two motifs special value, since they constitute ideal drug targets for structure-based drug design. The position of the two key residues (Arg393 and Phe444) is shown in figure 13. The side chains of the residues Arg^{393} is only 2.5 Å away from the phosphate backbone of the DNA. On the other hand the distance of Phe444 is almost 15 Å away from the phosphate backbone of the DNA. The DNA strand located in the channel of the HepC helicase (figure 13) was not certain if it was the translocating strand or the complementary strand. The interactions between the Arg393 residue and the oligonucleotide (figure 13) reveal that the binding is too specific and this implies that the strand in the channel of the helicase is probably the translocating one.

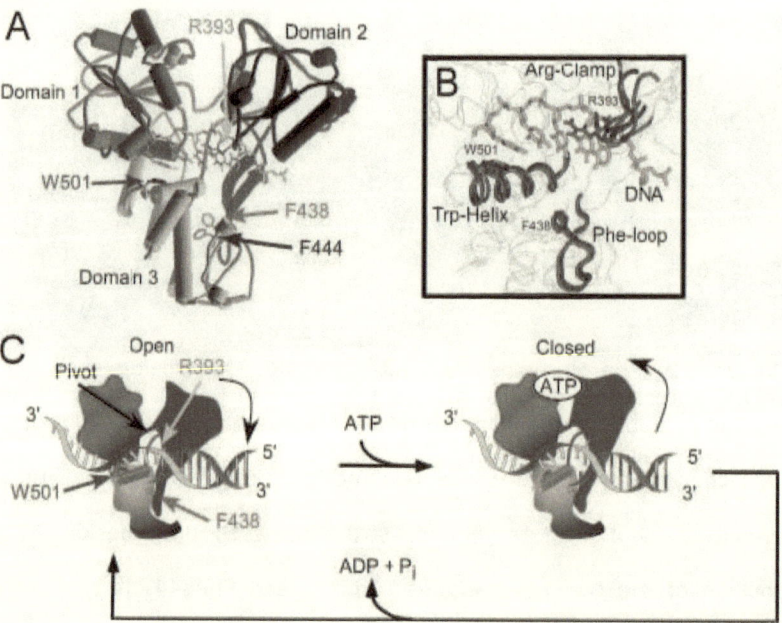

Figure 13. A and B show the locations of the Arg-clamp and the Phe-loop in the helicase structure of HCV. C is another attempt to explain the mechanism of action of the HCV helicase (Domain 1 is red, domain 2 is blue and domain 3 is gray). When ATP is not present the helicase is interacting with the oligonucleotide so strongly that it is impossible for it to slide through. After ATP binding though, Domains 1 and 2 come closer and the channel between domains 1-2 and 3 becomes bigger and this allows the oligonucleotide to slide towards the 3′ direction.

The helicase in figure 13 is probably in its ground-state configuration with the single-stranded oligonucleotide located in the major channel of the helicase protein. The oligonucleotide is peaking up a lot of interaction from the protein and this does not seem to favor the unwinding process. The problem is that the residues of the surrounding amino acids are too close to the ssDNA and this probably stabilizes it in the protein. The loosening of the interactions between the protein and the ssDNA is achieved by the contribution of ATP. ATP or any ATP analogs bind to the ATP binding site, which is on the same motif with the RNA binding site, the energy released by the hydrolysis to ADP will initiate an allosteric effect in the helicase that will eventually result in the movement of the second domain and the loosening of the interaction with the oligonucleotide. This will last until the effect from the hydrolysis of the ATP is finished. Then it the helicase will go back to its "closed" or "ground" state. To sum up it appears that offering energy to the helicase drives the protein to its "excite" state, and enables it to slide along the polynucleotide chain.

The sliding of the polynucleotide through the helicase follows a direction from 3' to 5' and never from 5' to 3' and this is achieved by translocating one strand of polynucleotide in turns. The helicase works as a dimer during unwinding. So, when the channel of one monomer is open, the other one is closed. Trp501 operates as a stop to the backsliding of ssDNA towards the 3' direction.

The effect of ATP on the helicase seems to be highly significant to the function of the helicase. The work from Yao et al. (6) illustrates just that. The structures of x-ray data of two helicases, one with the bound oligonucleotide and the other without it, were compared. Both structures came from the same HCV genotype and the sequence alignment proved that they are identical. It was proven that the helicase without the bound oligonucleotide is a bit more "closed" than the other one. Moreover it was found that in the helicase with the bound oligonucleotide, Arg393 is further away from the ssDNA when the Trp501 is further too. This proves that there is a coordinated action and movement of amino acids in the helicase during the oligonucleotide binding with a general loosening of interactions during the sliding f the helicases.

It is demonstrated in figure 13 that the Arg393 residue interacts with the bound ssDNA molecule both when ATP is present and not. It has also been shown that the ΔG of the DNA binding to the helicase in the presence of ADP is weaker when Arg393 is absent (mutated to Ala). Focusing on the residue 393 in the wild type helicase and in the mutated one (Arg393 to Ala393), the ΔG contribution of the 393 residue was found to be ten times stronger in the wild type one when the Arg393 is present. That is another prove of the fact that the Arg393 is an excellent target for drug design experiments.

The Phe-loop is basically focused on the amino acids Phe438 and Phe444, but it also involves the following three amino acids: His528, Phe531, Trp532 and Phe536. The vitality of the existence of these amino acids is confirmed by mutagenesis studies on the helicase. The mutation of any of those residues will lead to a mutant with no helicase unwinding capabilities.

Neighboring amino acids should also be considered. For example Arg393 next to the conserved Tyr392, which is thought to work in a similar function with Trp501. So, Tyr392 is supposed to hold nucleotide chain from backsliding towards the 3' direction. Mutating Tyr392 to Ala though does not change the unwinding capabilities of the helicase at all.

Another vital amino acid that was found after examining the binding mode of ssDNA in the HepC helicase is the Cys431. This amino acid is showed to have bonded with a small molecule, probably from the crystallization process. The capability of this residue to interact with a molecule alien to the protein-ssDNA complex is proof that Cys431 is good target for drug design. Furthermore the accessible area of the residue (to the solvent) is large (~75%) and the position of the residue in the helicase 's channel extremely strategic. As it will be described later on, the strategic positions of residues such as Arg393 and Cys431 will be exploited by structure-based drug design experiments in an attempt to synthesize compounds capable of establishing interaction with the above amino acids, thus blocking the way of the single strand of the polynucleotide.

1.4 Known compounds that work on the HepC Helicase

Research on the production of new drugs for Hepatitis C has been very slow, because of the lack of an efficient culture system of the virus. A lot of research has been focused on the NS3 domain of the Hepatitis C virus, because of its vital functions. The problem is that the majority of the drugs available are targeting the Protease protein of the NS3 domain and not the Helicase. Still some work has been done on the Helicase as well. The most important and most representative compounds designed for the HepC Helicase are described here:

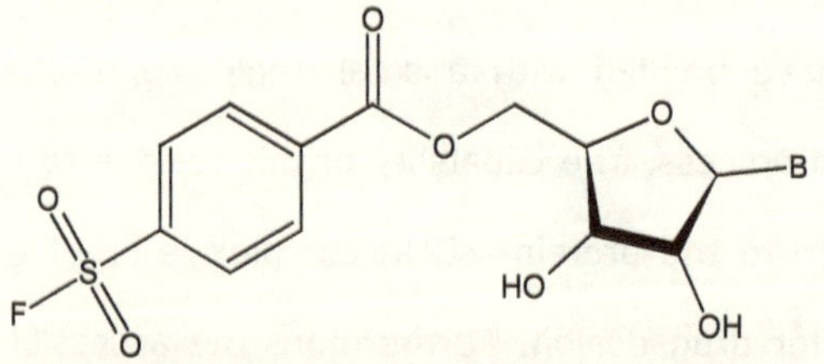

FSBR	= 5'-O-(4-fluorosulphonylbenzoyl)-esters of ribavirin	B = 1,2,4-triazole-3-carboxamide
FSBA	= 5'-O-(4-fluorosulphonylbenzoyl)-esters of adenosine	B = adenine
FSBG	= 5'-O-(4-fluorosulphonylbenzoyl)-esters of guanosine	B = guanine
FSBI	= 5'-O-(4-fluorosulphonylbenzoyl)-esters of inosine	B = hypoxanthine

Figure 14. 5'-O-(4-fluorosulphonylbenzoyl)-esters of ribavirin

5'-O-(4-fluorosulphonylbenzoyl)-esters of ribavirin, adenosine, guanosine and inosine were prepared and proposed as anti-viral agents against the flaviviridae family of viruses. The inhibition results were very moderate. Only in the case of the 5'-O-(4-fluorosulphonylbenzoyl)-esters of inosine there was some inhibition observed ($IC_{50} \geq 120$ µM). In the cases of FSBR and FSBA the inhibition levels were: 500 µM < IC_{50} < 1 mM. Surprisingly FSBG stimulated the function of the helicases. [99]

Virofarma reported a series of compounds with benzimidazole or benzimidazole-like moieties attached to a linker. The linker is usually pretty long and symmetrical. The IC50 values of these compounds ranges from 0.7 to 10 µM. The mechanism of action of these drugs is unclear, but based on their shape one could assume that they should act in the RNA biding site of the helicase, competing with the single-stranded RNA for binding to the protein.

Table 1. The 7 compounds reported by Virofarma and their activities.

Compound	Linker (R)	Linker (R) Name	IC_{50}
1		Benzene	10
2		A	0.7
3		C2	0.7
4		C4	0.7
5		C8	0.7
6		C10	0.7
7		C12	0.7

In total the Virofarma patent covers 66 compounds. The aminophenylbenzoxazoles- and the aminophenylbenzothiazoles-containing compounds showed no inhibitory effect, whereas the aminobenzimidazole-derived diamides produced a rather moderate 13% inhibition. The aminophenylbenzimidazolederived diureas showed inhibition in the range between 20 and 28%. In general it was noted that there is significant decrease of potency when the benzimidazole moiety (1a–e) is removed and substituted with benzoxazole or benzothiazole moieties. The same is for the removal of the benzene ring. Activity drops when the benzene ring is removed. SAR studies revealed that the NH group that is in the benzimidazole ring, the benzene group, located at the C-2 position of benzimidazole, and finally the nature of the linker are essential for inhibitory activity. A crude explanation of the importance of the NH group can be that the NH group could interact with the enzyme by H-bonding and the ring via hydrophobic interactions.

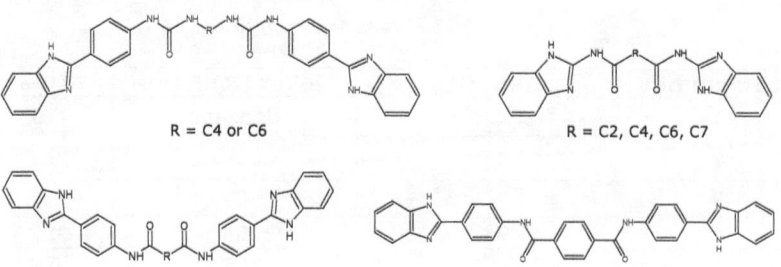

R = C4 or C6

R = C2, C4, C6, C7

R = C4 or C6

Figure 15. Some of the compounds suggested by Virofarma as HepC Helicase Inhibitors.

Novel piperidine derivatives were proposed as inhibitors for the HepC helicase by Vertex Pharmaceuticals.

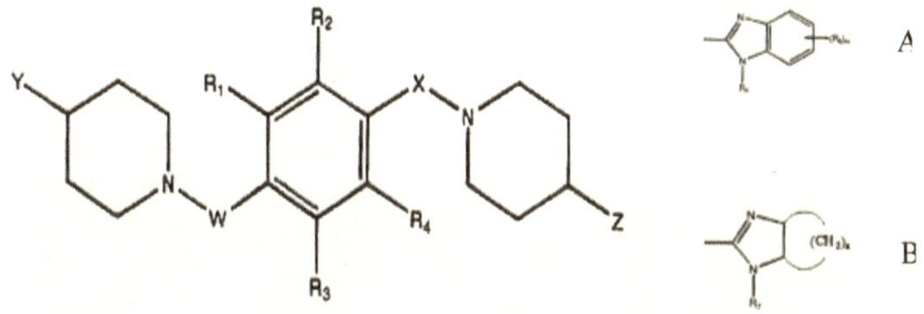

Figure 16. The generic formula of the compounds patented by Vertex Pharmaceuticals. A and B can take the places of Y and Z on the generic compound. The rest subsituents are summarized in table 2.

Table 2. The moieties table from Figure 16.

W, X	Alkylene (C_1- C_5) and carbonyl (-(C=O)-)
R_1, R_4, R_4, R_4, R_6	H, alkyl (C_1- C_6), halogen, hydroxy, alkoxy, carboxy, carbalkoxy, alkylthio, alkylsulfinyl, alkylsulfonyl, amino, acetamido, sulfonamido, alkylamino, dialkylamino and NO_2
R_5	H, alkyl (C_1- C_6), acyl
R_7	H, alkyl, acyl
n	3-5

Vertex produced a series of compounds capable of inhibiting the HepC helicase. These compounds include series of mono-, di-, and tri-substituted pentacycles. The pentacyclic ring may contain from one to three heteroatoms. The pentacyclic ring can also vary by being saturated, partially unsaturated or fully unsaturated.

Table 3. Full list with activities of the Vertex reported compounds (more details in Appendix 8).

Compound number	Compound structure ref.	Coupled NS3 ATPase (% inhibition)	HPLC ATPase IC_{50} (microM)
1	1	50 - 75	25 – 50
2	10	75 – 100	Less than 25
3	11	50 - 75	Less than 25
4	12	75 – 100	Less than 25
5	13	75 – 100	Less than 25
6	16	75 – 100	Less than 25
7	17	75 – 100	Less than 25
8	18	75 – 100	Less than 25
9	19	50 – 75	Less than 25
10	20	75 – 100	25 – 50
11	21	75 – 100	Less than 25
12	22	75 – 100	Less than 25
13	26	75 – 100	Less than 25
14	30	75 – 100	Less than 25
15	31	75 – 100	Less than 25
16	32	75 – 100	Less than 25
17	34	50 – 75	25 – 50
18	35	75 – 100	Less than 25

According to the work of Morris et al [79] potential inhibition of the NS3 helicase can be achieved by intracellularly expressed antibodies. So, antibody fragments against HCV helicase were prepared in an attempt to stop the HCV replication. A series of human immunoglobulin genes were obtained and high affinity antibodies were selected. Their specificity against NS3 helicase was confirmed by a helicase-mediated DNA unwinding Assay that was developed in ELISA-format. Eventually a set of high affinity antibody fragments specific for the Helicase of HepC were developed and isolated. The results of their application have shown complete and full inhibition of the helicase activity of the NS3 domain. So, a new approach for gene therapy and better application of the antibodies is currently being considered.

A series of nucleotide compounds were suggested by Borowski et al. [82] as potential inhibitors of the helicase of HepC mainly targeting the ATP site on the protein. The theory is that since the unwinding process is energy-depended, even competition for the ATP, induced by any compound, would result in less energy being available for the system and as a result reduced unwinding function of the helicase. So, a wide range of competitive NTPase inhibitors was suggested. Those include ribavirin-5 triphosphate (RTP), ribavirin-5 diphosphate (RDP), adenosine-5-g-thiotriphosphate (ATP-g- S) or ADP. These compounds were all tested against HepC helicase and were found to be moderate inhibitors of the unwinding activity of the HCV

Enzyme (summarized in figure 17).

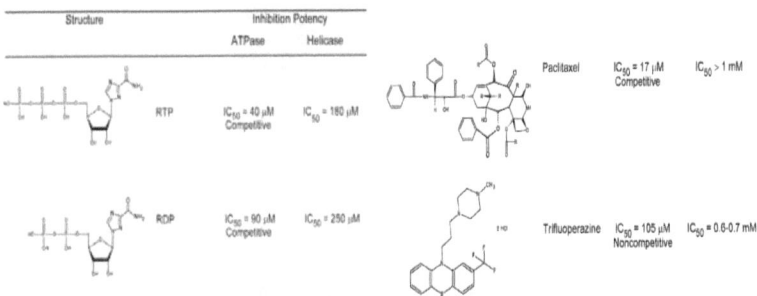

Figure 17. The series of compounds suggested by Borowski [82]

A relative to the antracycline antibiotics (Zhu et al., 1999; Bachuret et al., 1992) is the mitoxantrone. The mitoxantrone was tested and managed to reduce the helicase activity of the HCV NTPase/helicase. Of course, these compounds are still far away from being drugs because of their high cytotoxicity and low bioavailability. Nevertheless, it is a promising series of compounds that need to be optimized in an attempt to keep their helicase inhibitory properties while reducing their toxicity and increasing their bioavailability (figure 18).

Inhibitor	Structure	Potency
Doxorubicin	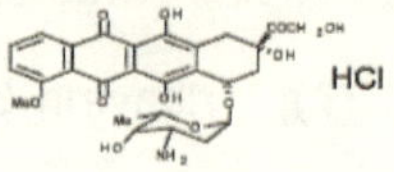	$IC_{50} = 5.0 \ \mu M$
Daunomycin	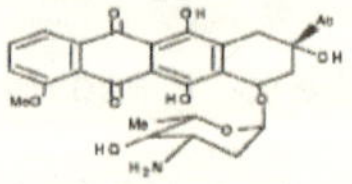	$IC_{50} = 57 \ \mu M$
Epirubicin	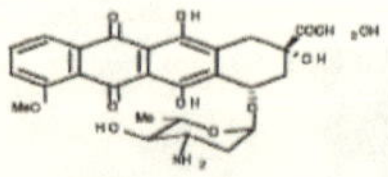	$IC_{50} = 0.75 \ \mu M$
Nogalamycin	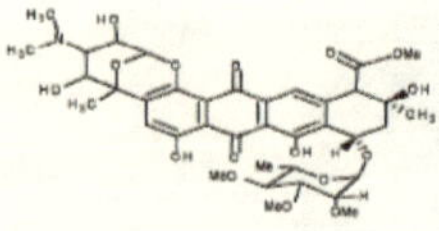	$IC_{50} = 0.1 \ \mu M$
Mitoxantrone	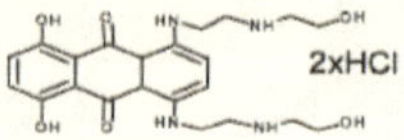	$IC_{50} = 6.7 \ \mu M$

Figure 18. The list of the the antracycline antibiotics (Zhu et al., 1999; Bachuret et al., 1992)

The effect of N7-chloroethylguanine and N9-chloroethylguanine in the helicase activity of the WNV was investigated. These compounds were found to be activators of the helicase activity by 850 and 220% respectively (Borowski, 2001a). Similar patterns in activation were observed with the HepC helicase as well (figure 19).

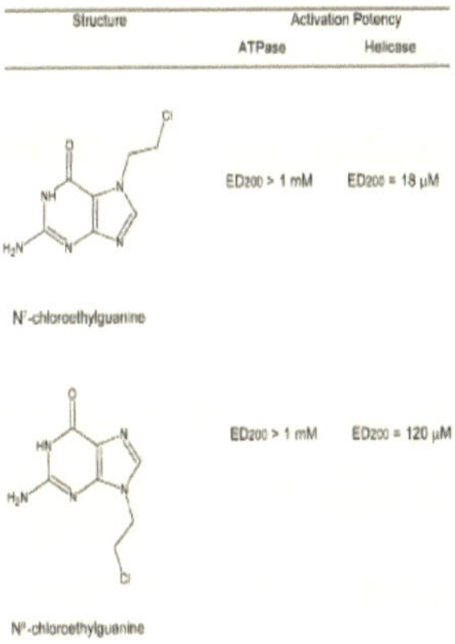

A series of compounds identified by

Structure	Activation Potency	
	ATPase	Helicase
N7-chloroethylguanine	ED200 > 1 mM	ED200 = 18 µM
N9-chloroethylguanine	ED200 > 1 mM	ED200 = 120 µM

Figure 19. N7 chloroethyl guanine and N9-chloroethyl guanine

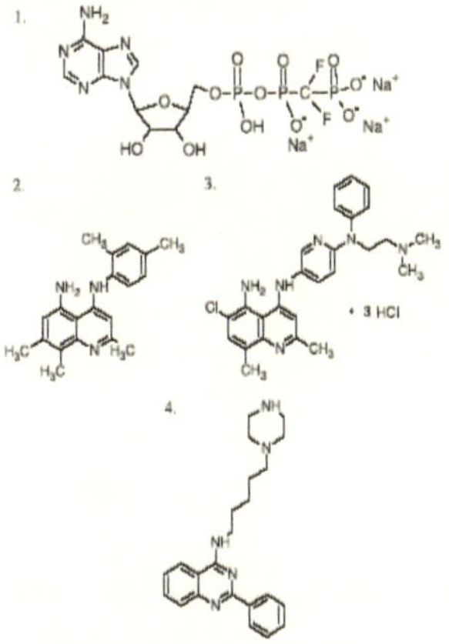

Fig. 1. Structures and chemical names of compounds examined: (1) β, γ-difluoromethyleneadenosine 5'-triphosphate (AMP-PCF₂P), (2) 4-(2, 4-dimethylphenyl)-2,7,8-trimethyl-4,5-quinolinediamine, (3) 6-chloro-N-4-(6-[[2-(dimethylamino)ethyl](phenyl)amino]pyridin-3-yl)-2,8-dimethylquinoline-4,5-diamine-HCl, and (4) 2-phenyl-N-(5-piperazin-1-yl-pentyl)quinazolin-4-amine.

HTS are shown in the left. The nonhydrolyzable ATP analog, b; c – difluoromethylene - adenosine – 50 - triphosphate (AMP-PCF2P) was also used. HTS studies indicated that the compounds 4 - (2,4 - dimethylphenyl) - 2,7,8 - trimethyl-4,5-quinolinediamine, DAPI (40,6 – diamidino - 2-phenylindole dihydrochloride hydrate), AMP-PNP, and Chaps to be used as potential inhibitors for the HepC helicase.

A series of ring-expanded ("fat") nucleoside analogues with the 6-aminoimidazo [4,5-e] [1,3] diazepine-4,8-dione ring system was synthesized and tested against the NTPase/helicase WNV. Some compounds were found to be potent inhibitors of the NTPase/helicase of the WNV. It was also found that the same compound that was active against the helicase activity of the protein was inactive against the NTPase activity. It is supposed that these series of compounds will bind to the major or minor groove of dsDNA or dsRNA and affect the stability of the double helix.

Table 4. The IC50 values of the above compounds

Compound Number	Sited number of compound	WNV IC$_{50}$ (µM)
1	16	10
2	30	1,5
3	39	1-3

Two series of ring-expanded ("fat") heterocycles, nucleoside and nucleotide analogues were prepared and tested against the NTPase/helicase activity of the Flaviviridae family of viruses. The first series was containing the imidazo [4, 5-e][1,3] diazepine ring system and the other series containing the imidazo [4,5-e][1,2,4] triazepine ring system. They were tested against the helicases of WNV, HCV and JEV. It was found that some of these compounds were quite potent inhibitors of the corresponding helicases that they were tested with. Their inhibitory effect is mainly exerted on the ATPase activity of the enzymes [82].

1   2

Table 5. The IC50 values of the above compounds

Compound Number	Sited no. of compound	WNV IC$_{50}$ (μM)	WNV IC$_{50}$ (μM)	WNV IC$_{50}$ (μM)
1	15	5.7	>500	2
2	24	3.3	5.5	>500

CHAPTER 2:

Molecular Modelling

2.1   Introduction

The biological and chemical activities of a compound in relation with
their corresponding active site on a protein constitute the ultimate
target for molecular modelling prediction. Molecular modelling
techniques have been made and improved in the search for the
accurate prediction tool of protein (active site) – compound
relationships. There are two mainstream approaches to protein-
compound interaction prediction. First is the "no lead" approach,
where prediction of the cooperation of a certain complex is based on
chemical rules and chemical properties of the involved parties,
whereas in the second approach either a natural inhibitor or a series of
active compounds have been reported to act on a certain active site.
So, properties of known drugs have led to the establishment of several
techniques such as structure-activity relationships (SAR) that may help
in the prediction of more active compounds. These experiments have
been optimized by simulation techniques such as like QSARs and
CoMFAs.

The HepC Helicase (pdb code: 1A1V) is a difficult receptor for drug
design. It does not have a specific site to be targeted, except from the
ATP site. So, quite normally some of the research done on the helicase
has been focused on the ATP site, involving the docking of ATP and

ATP analogues. The main aim of the modelling part though is to design compounds that will act on the RNA binding motif of the helicase, they will establish interactions with residues on it thus preventing the sliding of the ssRNA through the helicase. The size of the RNA binding motif though is too large for small molecules to work on it. So, all the residues that could potentially contribute to interactions with compounds are isolated and designated as target residues.

2.2   The Hepatitis C Helicase Protein

The Hepatitis C Helicase was extensively studied both with and without a nucleic acid stand docked in the enzyme's channel. In fact a variety of different structures were studied but the main focus was on the HepC with the nucleic acid present in the RNA binding channel. The structures that were used in this study are summarised in the table below (table 6).

Table 6. The different structures of the HepC Helicase. All of the structures below were obtained from the Expasy-pdb database.

Structure ID	PDB code	Complexes
Hepatitis C Helicase NS3	1A1V	Monomer+DNA
Hepatitis C Helicase NS3	8OHM	Monomer
Hepatitis C Helicase NS3	IEH1	Dimer

All the sequences of the various HepC Helicases were imported into the LPC analysis software. LPC analysis software (Ligand - Protein

Contacts analysis software) was used for the active site analysis studies. LPC automatically: 1. identifies the residues that make the active site of the protein; 2. calculates the solvent accessible surface of every atom; 3. determines the contacting residues and type of interaction they undergo (hydrophobic-hydrophobic, aromatic-aromatic, etc.); 4. indicates all putative hydrogen bonds. LPC also predicts changes in binding strength following chemical modification of the ligand. 5. calculates all the distances between the ligand and the protein's active site.

Table 7. LPC ANALYSIS OF THE LIGAND CONTACTS OF 1A1V.

```
--------------------------------1A1V------------------------------
                                ----------------------------------

     Residue        Dist    Surf    HB    Arom    Phob    DC
   369A  HIS*        4.5     4.7    -      -       -       -
   412A  ASP*        3.9    16.2    +      -       -       +
   430A  THR*        1.3    76.7    -      -       -       +
   432A  VAL*        1.3    76.1    +      -       -       +
   433A  THR*        3.9    15.8    +      -       -       +
   449A  THR         3.5     0.9    -      -       -       +
   450A  THR*        3.4     3.0    -      -       -       -
   451A  LEU*        2.7    44.7    +      -       -       +
   452A  PRO         4.1     3.4    -      -       -       -
   453A  GLN*        3.7    13.2    -      -       -       -
   454A  ASP*        3.6    17.9    -      -       -       +
   457A  SER*        3.5    29.6    -      -       -       +
   461A  ARG*        4.2    10.5    -      -       -       +
   481A  ARG*        4.1    11.7    -      -       -       +
     4B   U*         5.7     2.7    -      -       -       -
-----------------------------------------------------------------
 Legend:
Dist - nearest distance (Å) between atoms of the ligand and the residue
Surf - contact surface area (Å²) between the ligand and the residue
HB    - hydrophilic-hydrophilic contact (hydrogen bond)
Arom - aromatic-aromatic contact
Phob - hydrophobic-hydrophobic contact
DC    - hydrophobic-hydrophilic contact (destabilizing contact)
+/-   - indicates presence/absence of a specific contacts
*     - indicates residues contacting ligand by their side chain

    → FULL TABLES OF THE LPC ANALYSIS ARE SUMMARISED IN APPENDICES 6-9
```

The second software package that was used is the CSU program (Contacts of Structural Units). The CSU program calculates contacts of structural units such as helices, sheets, strands and residues. Both approaches are based on a detailed analysis of inter-atomic contacts and interface complementarity.

LPC analysis of every non-protein structure can give us information on the accessibility and availability of certain residues. It is possible to determine that Cysteine 431 and Valine 432 are quite accessible to solvent and willing to get involved in interactions with potential compounds that could be targeting their area. Table 7 summarizes the LPC analysis for the 1A1V helicase. In such a table it is possible to identify the residues that make the active site of each protein-ligand complex, the accessible surface, of each residue to the solvent, the distance of the ligand from the residue and the type of specific and the non-specific interactions between the ligand and the various structures that surround it.

The first task was to prepare the files so that they would be suitable for further molecular modelling experiments. In the case of 1A1V, there ware two major problems. The first was the presence of a single strand of DNA through the helicase domain, and the second one was the presence of a little molecule, probably leftover from the crystallization of the protein, which had established an S-S bond, a sulphur bridge, with Cysteine 431. The presence of these two alien structures did not mean that the protein structure was not suitable for experimental use, but it definitely meant that a few preparatory steps had to be taken. Both the DNA and the small compound attached to the Cysteine were manually removed and taken out of the protein structure, by editing the PDB file. The protein's amino-acid conformation was still unchanged, as if nothing had been removed. The protein had to be changed from its tense condition to its relaxed condition. The bare protein structure was loaded to a variety of different software packages in order to energetically minimise it and bring the structure to its relaxed state.

Energy minimization alters molecular geometry to lower the energy of the system, and yields a more stable conformation. As the minimization progresses, it searches for a molecular structure in which the energy does not change with infinitesimal changes in geometry. This means that the derivative of the energy with respect to all Cartesian coordinates, called the gradient, is near zero. This is known as a stationary point on the potential energy surface. If small changes in geometric parameters raise the energy of the molecule, the conformation is relatively stable, and is referred to as a minimum. If the energy lowers by small changes in one or more dimensions, but not in all dimensions, it is a saddle point. A molecular system can have many minima. The one with the lowest energy is called the global minimum and the rest are referred to as local minima.

MM+ is unique among the force fields in the way it treats bonds and angles. Both the bond and angle terms can contain higher order terms than the standard quadratic. PM3 on the other hand is based on the neglect of diatomic differential overlap (NDDO) approximation. NDDO retains all one-center differential overlap terms when Coulomb and exchange integrals are computed. The parameters for PM3 were derived by comparing a much larger number and wider variety of experimental versus computed molecular properties. That's why PM3 is known as a semi-empirical optimisation technique.

Superimposing the initial structure of the helicase with the ssDNA and the small compound removed with the energetically minimised structure, produced a RMSd of 0.77 Å. Although the RMSd variation is lower than even one angstrom overall, it could mean that at some points across the protein the RMSd between specific residues could have been a lot higher. The RMSd of 0.77 is the mean RMS of the whole protein. The figure below shows the superimposition of the tense and relaxed state of the hepatitis NS3 domain (1A1V). The tense state is in green and the relaxed in red.

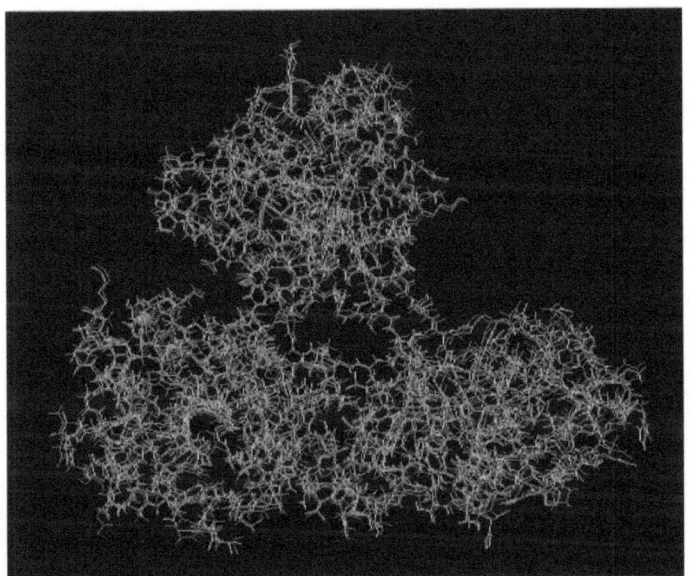

Figure 20. Superimposition of the tense and relaxed states of 1A1V

Below there is an interaction map between the nucleic acid chain and the protein of helicase NS3 domain made with the aid of LigPlot (figure 21). More detailed LigPlot interaction maps can be found in Appendix 3. Figure 24 contains images of the dockings of single and double stranded RNA. The double stranded RNA has not been properly docked since there is not a point to it. The point in fitting dsRNA through the channel is to prove that the channel is huge. However, the ssRNA through the channel and the other helix under the NS3 domain scenario has been docked. It can be seen from the last image that both the two strands fit nicely in the channels in-between domains 1-2 and 3 and under domains 1 and 3. It does not mean that this is the splitting mechanism of dsRNA on the helicase but it is an approach that could be further investigated. According to the limited literature available the splitting of the dsRNA does not take place on the NS3 domain, but before that. It has been suggested that RNA reaches the helicase as a single strand and not as dsRNA [14].

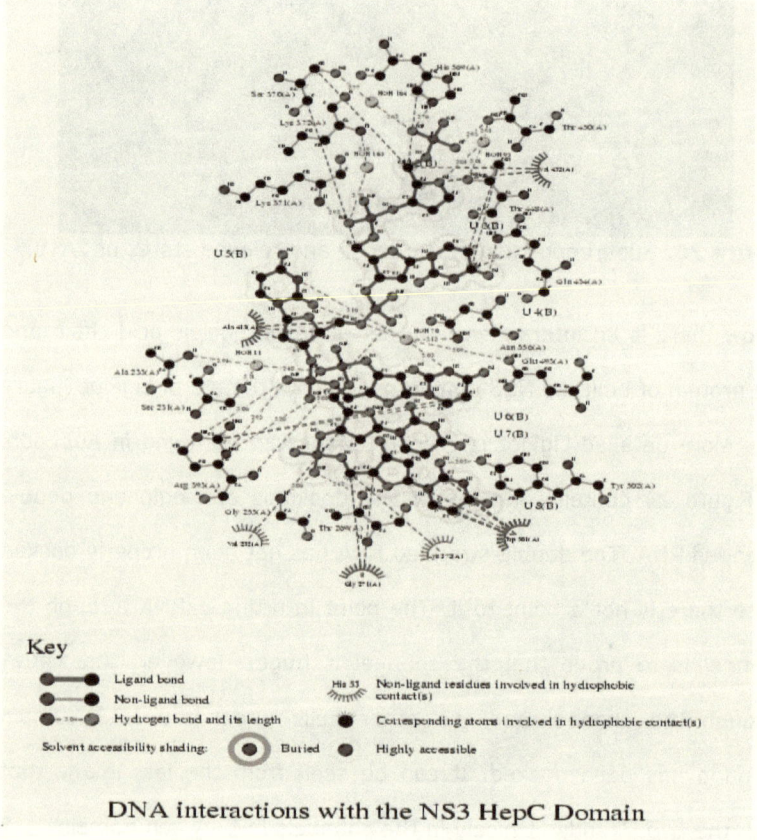

DNA interactions with the NS3 HepC Domain

Figure 21. Interaction map between the ssRNA and the helicase

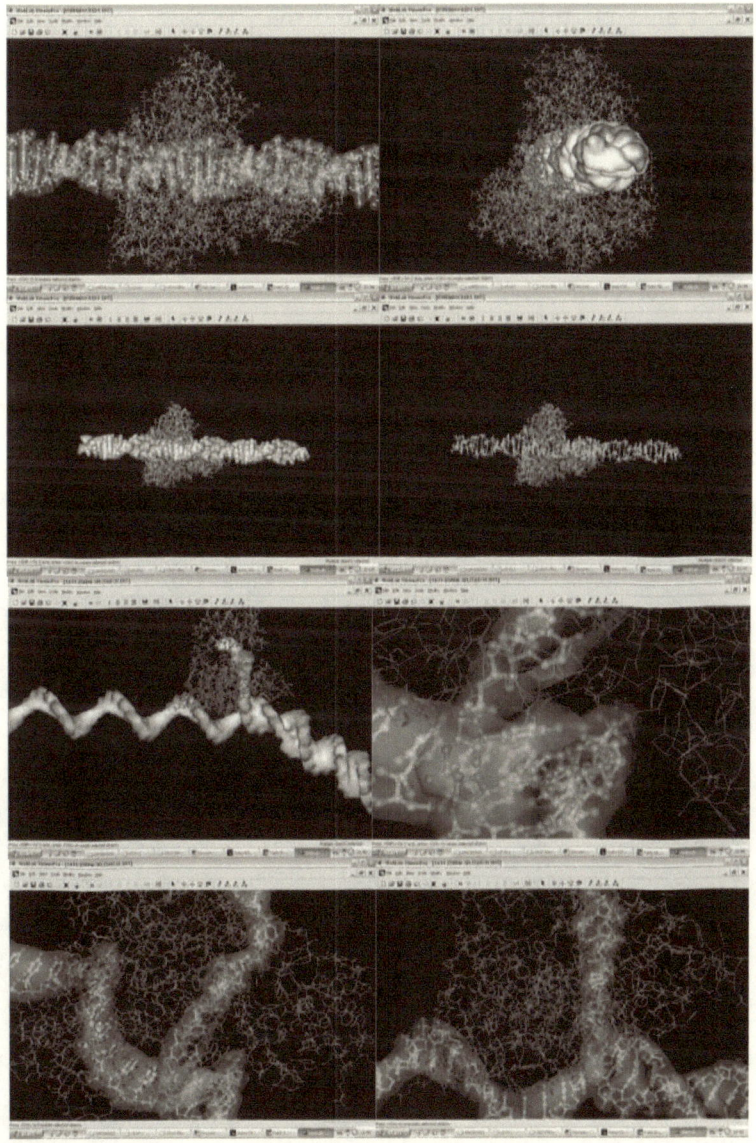

Figure 22. Docking of ssRNA and dsRNA in the channel of the HCV helicase

Final stage of the experiments on the helicase involved the investigation of the protein for another potential active site available for possible future targeting. A six-dimensional search was performed on the protein and all the cavities were revealed. It was found that there is a site between the domains 1 and 3 that could be possibly targeted for drug design. The site is shown in figure 23. The drug design techniques used here were exactly the same as the ones described above. The series of the new compounds designed for that active site on the helicase are listed in figure 36.

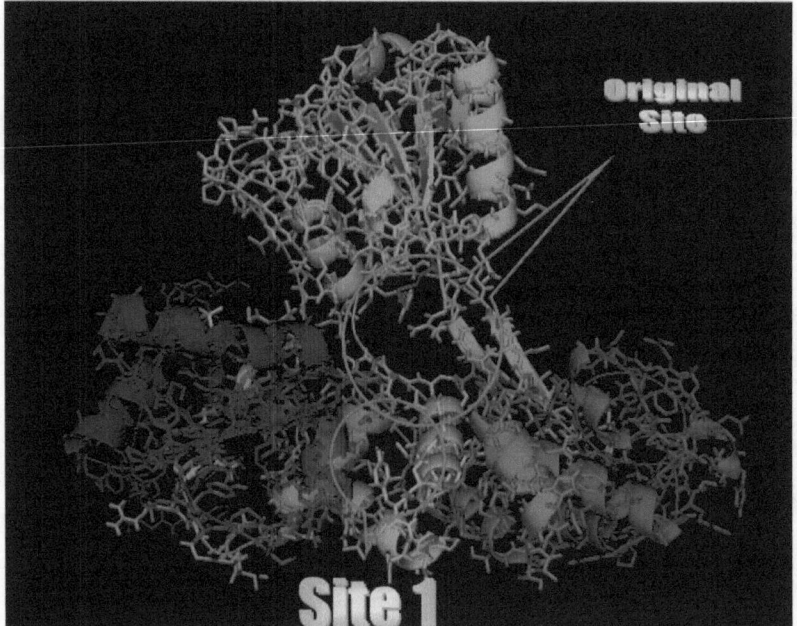

Figure 23. The original and the new (site 1) sites for possible inhibition on the helicase

CHAPTER 3:

Homology Modelling

3.1   <u>Introduction</u>

The aim of this project is to design the 3D structures of the Helicases and Polymerases of the most important members of the Flavi virus family out of known crystal structures, commonly of the hepatitis C, which has been extensively studied over a long period of time. The members of the Flavi family that are of high interested and have been studied in this chapter are the Dengue virus, the West Nile virus, the Japanese Encephalitis virus and the Yellow Fever virus. Both the Helicase and the Polymerase protein model series was constructed for the above species using primarily the Hepatitis C Helicase and Polymerase x-ray resolved 3D structures as templates.

There are two ways to build a 3D model by homology modelling. The first way is based on strict superimposition. The second method is based in geometric restraints, which includes analogy to the molecular replacement and distance geometry methodologies. Composer takes into account both of them. There are a number of algorithms that have been developed in order to help make a model. The most popular ones are the Composer, Jig-Saw and Modeller. In this experiment both the Modeller and Composer were used with the latter being a bit more efficient at building the SCRs. Usually the construction of SCRs uses similar SCRs in sequence from proteins that are homologous to each other. That is done in order to define a new core out of a multiple alignment. So, first a framework from SCRs is being made by superimposing known structures. Then the chosen homologue sequence is superimposed on the unknown sequence in the main chain of the framework, which has been averaged. Then the anchored stumps on the framework are found and compatible structures are placed on them. Finally, the side chains are modelled. Now, the quality of the models had to be checked and confirmed.

The N-CA and the CA-A bonds in a protein are capable of rotating. The rotation of these bonds can be described by the phi and psi torsion angles. Little protein models derived from computers can be used to manipulate those phi and psi torsion in search of more stable conformations [GN Ramachandran]. Each structure is being investigated for atom-atom close contacts for every possible conformation. The atoms are not treated as dots in space anymore, but as spheres with a radius equal to each atom's van der Waals radius. Those spheres are not flexible, so if the phi/psi configuration of a protein makes two spheres to collide (i.e. occupy common space), then that would result in a sterically disallowed conformation of the backbone of the protein.

3.2 Homology Modelling of the Helicase Protein Model Series of the Flavi family.

The sequence alignment was done with a multiple alignment program called ClustalW (Higgins and Sharp, 1989). The sequences had to be in FASTA format to be able to be used as input for the program. The whole process of the sequence alignment is broken down into four steps. Firstly, using rapid alignment methods the program calculates all pairwise similarity scores. The second step is the generation of a similarity matrix. Then the sequences are clustered according to the generated similarity matrix with the aid of an algorithm. Next step is the making of a cluster alignment using a consensus method and finally a multiple-progressive alignment is generated. The groups of the sequences are aligned according to their cluster branch order.

The primary sequences of the NS3 domains of the Dengue virus, the West Nile virus, the Japanese Encephalitis virus and the Yellow Fever virus were obtained from GenBank. These sequences were then aligned with the known HepC sequences and the protease part of the NS3 domain was removed. The remaining sequence was the Helicase one. The Helicase sequences were then aligned against all the available known helicase sequences obtained from the BrookHaven Database, and the ones with the best scores were picked out to be used as templates during the homology modelling. It was found that the best alignment score was achieved with the HepC helicase (1A1V & 8OHM).

The models were obtained from COMPOSER, a module of the Tripos Sybyl suite running on a SGI workstation. Homology modelling is the comparison of a protein sequence of unknown structure with that of a protein with known structure, in order to predict what the unknown structure should, ideally, look like. The primary sequences of each of the unknown structures were imported in COMPOSER and the sequence of the corresponding template proteins were added in the database of the software. The template had to be in PDB format, and to be added in the existing database of COMPOSER. As soon as that was done, the program would search for homologues in its database and the imported sequence was found and selected. Then COMPOSER did a sequence alignment of the two sequences. Initially, there were 12 loops for the Dengue virus model, 16 for the West Nile virus, 22 for the Japanese Encephalitis virus and 31 for the Yellow Fever virus. Composer though managed to fix all of them using its database. The database of COMPOSER at that time numbered 1850 structures. In order to assist the database of COMPOSER to become more efficient with the two helicases, a group of 125 helicases was downloaded and imported into the database. The templates would offer the SCRs and the rest of the database (with major contributors the group of the Helicases) the SVRs. Finally all loops were eliminated and the four models for the four Flavi virus members were built.

The best sequence alignment for the Helicase models of the Dengue and the West Nile viruses was the 1A1Va version of the HepC helicase, whereas for the Japanese Encephalitis and the Yellow Fever Helicase models the 8OHM version of the HepC Helicase was used. The minimum percentage identity of two structures has got to be 30% or if that is not feasible as close as possible to 30 percent. If the percentage identity is below 30%, then the two proteins are not considered to be similar and as a result, traditional homology modelling is not possible. Similarity above 30% means that the two sequences are similar enough in their primary sequences and as a result, they could possibly share similar structural characteristics as well. The identity scores for the all the helicase models was marginal, but still good enough to be used for further experimentation of homology modelling. The identity percentage between the Dengue Primary sequence and the 1A1Va HepC sequence was 31.2%, the West Nile virus 29,8%, the Japanese Encephalitis 30% and the Yellow Fever 31%. Things with the Polymerase project though were not as straightforward as with the helicase one. The average initial identity between the primary sequences of the four species against the HepC template was approximately 17% and this made the task of building the models quite more complex.

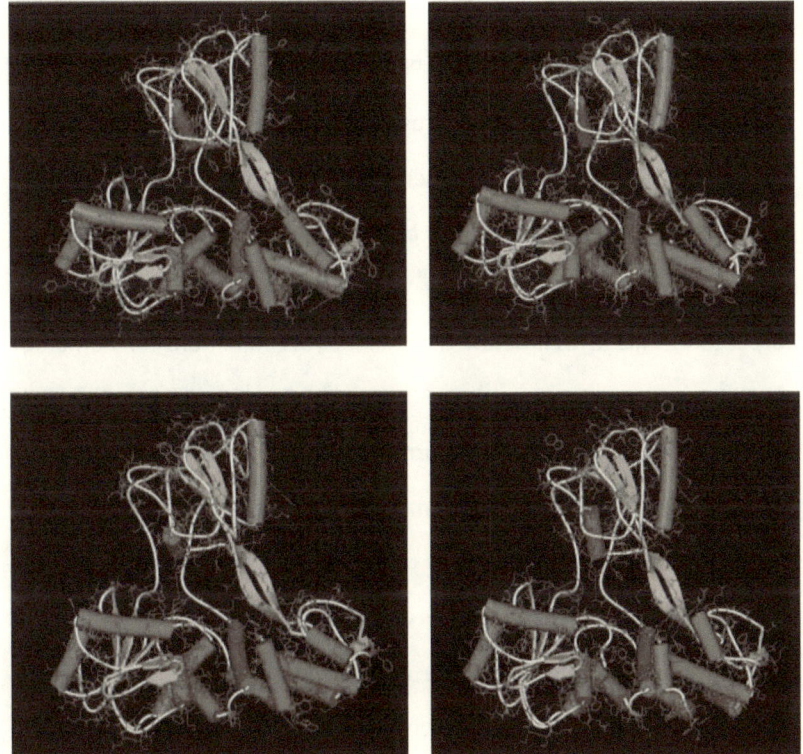

Figure 24. The four Helicase Models: the Dengue virus (top left), the West Nile virus (top right), the Japanese Encephalitis virus (bottom left) and the Yellow Fever virus (bottom right).

3.3 Homology Modelling of the Polymerase Protein Model Series of the Flavi family.

The theoretical structure of the Polymerase proteins for each of the species of Dengue, West Nile Virus, Japanese encephalitis and yellow fever were resolved using the Hepatitis C polymerase as a template structure. The homology identity between the Hepatitis C polymerase and the primary sequences of the above four species is approximately 20 percent. This is well below the nominal set minimum for homology modelling experiments, which is 30 percent. Still, homology is possible and can give conclusions and information that are vital for drug design experiments. The sequence alignment between the four species above (figure 24) and the four species with the template reveals four key residue "locations" that are conserved and play key roles in interacting with the ssRNA, as found in the crystalized Hepatitis C polymerase. Those four residues are in accordance with the existing literature, where they have been characterized as key residues for the interaction between the protein and the oligo RNA. The whole homology modelling process was focused on these residues, the location of the ssRNA and the shape of the Hepatitis C Polymerase. Involving information such as the interaction of the polymerase with RNA, known from the Hepatitis C polymerase, which has been co-crystalized with a ssRNA in the major channel, the shape of the Hepatitis C polymerase and the aid-confirmation of secondary structure prediction algorithms will compensate the low starting homology identity between the four species of unknown structures and their template (the Hepatitis C Polymerase). The purpose of homology modelling of species within the same family of the Hepatitis C polymerase is to identify the relative positioning of the residues of the active site of each protein, that are more likely to interact with the oligo RNA. So, successful homology modelling in the active site region is all that is needed. The precision and the success of the homology modelling in the outer regions of the models is not considered to be as important. The fact that the template used here has been co-crystalized with a ssRNA in it helped a lot, since the coordinates of the ssRNA was used as a scaffold around which the protein will have to be build. Retaining the coordinates of the ssRNA and the coordinates of the residues that establish the key interactions with the RNA, while being conserved within the Flavi family compensates the low starting homology identity. The shape of the Hepatitis C polymerase was also used as a criteria to build the new models. The ssRNA was removed from the Hepatitis C structure. The structure was then hydrogen fixed, charges were applied and finally it was minimized energetically. A single layer of dummy-inert atoms was laid over the relaxed structure of the Hepatitis C polymerase and the coat (balloon) that came out was used to define the shape and size of the models. Finally, putting together information such as the size, the shape and the positioning of the ssRNA and the key corresponding residues give enough data to proceed to a successful homology modelling experiment, good enough to give clues and information of the structures of the four unknown species of the Flavi family. The models will be used for screening purposes of molecular databases, and the site of interest as a target site for potential drugs is the RNA binding site region that has high probability to be of a scientifically viable prediction standards.

Figure 25. The superimposition of the key residues that interact with the ssRNA. The Arginine residue is located at the exact same position, with the exact same geometry and orientation on all models as in the template protein. The type of interaction with the ssRNA is identical with the template.

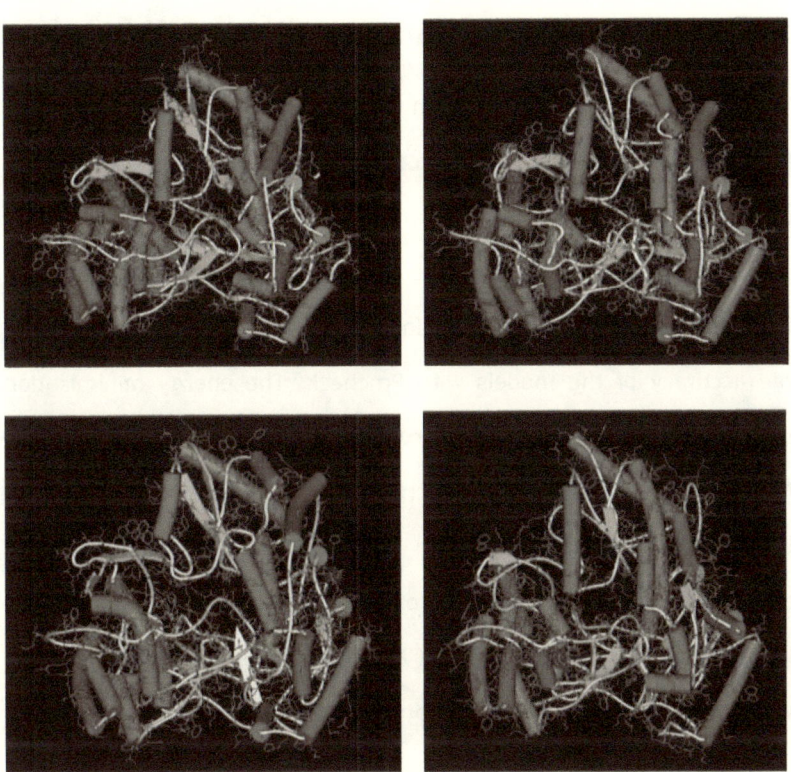

Figure 26. The four Helicase Models: the Dengue virus (top left), the West Nile virus (top right), the Japanese Encephalitis virus (bottom left) and the Yellow Fever virus (bottom right).

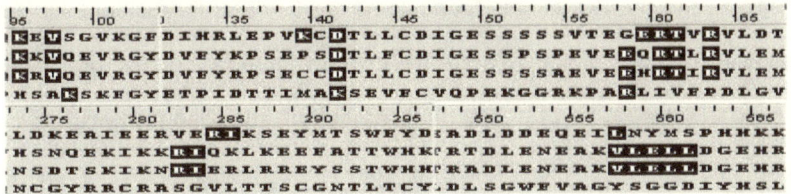

Figure 27. Example of the Sequence Alignment between the four species. In black squares are the amino acids that are closer than 4.5 A from the RNA backbone. There are a lot of concerved residues that interact with RNA (or at least are in close proximity) thus improving the quality of the models.

3.4   Model Evaluation & Optimization

The accuracy of the models from the homology modelling study was tested with PROCHECK. Before testing the models for their accuracy with PROCHECK, the templates had to be tested as a reference. Poor quality templates are not capable of producing viable models via homology modelling. Energy minimisations were done prior to testing the accuracy of the models with Procheck. The energy minimisation was initially performed in Sybyl running on an SGI using a universal MM algorithm and then with HyperChem on a PC running Windows XP, using firstly the MM+ algorithm and then the PM3 semi-empirical algorithm. The final step of optimisations was performed with HyperChem using the same set of algorithms, only this time the model was put in a water box and the forces of the water solvent were taken into account.

The 1A1Va model was found to have all its residues in the allowed regions and the 8OHM was found to have 99.7% of its residues in the allowed regions, which is accurate enough for a homology modelling study (tables 9 & 11). The first model of the Dengue virus was obtained with COMPOSER using the HepC Helicase 1A1Va as a template structure. The loops that COMPOSER found were automatically eliminated using COMPOSER's database of proteins and the extra set of Helicases that were manually imported into the

COMPOSER's database. When COMPOSER was finished the model was energetically minimised using Sybyl's molecular mechanics algorithm for 1000 iterations. The model was tested at this point with PROCHECK and it was found that 82.2 % of its amino acids were in the allowable area of the plot (RP), 13.9% were in the additional allowed regions and 2.5% were in the generously allowed regions, whereas 1.4% of the residues of the model were found to be in the disallowed area of the Ramachandran Plot (table 9). The total energy of the Dengue model after the initial minimisation from Sybyl was 26,500 Kcal, but after the further energy minimisations the total energies had dropped down to -2103 Kcal. This huge drop in energy meant that there were chemical errors in the model's structure that were contributing in an increase in the total energy of the model, thus making the model very unstable. Using more sophisticated energy minimization algorithms those errors were fixed and the system was made more stable. The increased stability of the system and the elimination of various errors on the structure of the two models were confirmed by PROCHECK. The percentage of residues in the allowed area of the RP, representing residues with non-conflicting atoms in their structures, increased to 92.5%, 5.3% in the additional allowed regions and 2.2% in the generously allowed area. There were no residues in the disallowed area (table 10).

Table 9. Ramachandran Plot statistics for HepC Helicase (1A1Va) & Dengue Virus (Model)
Ramachandran Plot statistics for HepC Helicase (1A1Va)

		No. of residues	%-tage
Most favoured regions	[A,B,L]	316	89.5%*
Additional allowed regions	[a,b,l,p]	37	10.5%
Generously allowed regions	[~a,~b,~l,~p]	0	.0%
Disallowed regions	[XX]	0	.0%
		----	------
Non-glycine and non-proline residues		353	100.0%
End-residues (excl. Gly and Pro)		7	
Glycine residues		38	
Proline residues		31	
Total number of residues		429	

Ramachandran Plot statistics for Dengue Virus (Model)

		No. of residues	%-tage
Most favoured regions	[A,B,L]	296	82.2%*
Additional allowed regions	[a,b,l,p]	50	13.9%
Generously allowed regions	[~a,~b,~l,~p]	9	2.5%
Disallowed regions	[XX]	5	1.4%*
		----	------
Non-glycine and non-proline residues		360	100.0%
End-residues (excl. Gly and Pro)		2	
Glycine residues		40	
Proline residues		27	
Total number of residues		429	

Table 10. Ramachandran Plot statistics for the improved Dengue Model

Ramachandran Plot statistics for Dengue Virus (Model)

		No. of residues	%-tage
Most favoured regions	[A,B,L]	333	92.5%
Additional allowed regions	[a,b,l,p]	19	5.3%
Generously allowed regions	[~a,~b,~l,~p]	8	2.2%
Disallowed regions	[XX]	0	0%
		----	------
Non-glycine and non-proline residues		360	100.0%
End-residues (excl. Gly and Pro)		2	
Glycine residues		40	
Proline residues		27	

Total number of residues		429	

Based on an analysis of **118** structures of resolution of at least **2.0** Angstroms and *R*-factor no greater than **20.0** a good quality model would be expected to have over **90%** in the most favoured regions [A,B,L].

The results for the accuracy of the model of the WNV showed that 80.9% of the model's residues were found in the core region of the RP, 13.6% were in the additional allowed regions, 3.7% of the residues were found in the generously allowed area of the RP and 1.8% of the model's residues were located in the disallowed region (table 11). The homology modelling procedure was repeated from the beginning once more in order to ensure that COMPOSER's run was flawless. The results obtained from the repeat of the experiment were completely identical to those of the first experiment. So, the only way to improve the quality of the models was to make sure that they are chemically accurate. This was done by further energy minimisations using a variety of different programs, algorithms and environments. The total energy of the model after the initial minimisation from Sybyl was 21,220 Kcal for the WNV. After the further energy minimisations the total energy had dropped down to -3105 Kcal for the WNV model.

Table 11. Ramachandran Plot statistics for HepC Helicase (8OHM) & West Nile Virus (Model)

```
┌─────────────────────────────────────────────────────────────────────────┐
│ Ramachandran Plot statistics for the HepC Helicase (8OHM)                 │
│                                      No. of residues      %-tage          │
│ Most favoured regions       [A,B,L]        323           88.5%*           │
│ Additional allowed regions  [a,b,l,p]       39           10.7%            │
│ Generously allowed regions  [~a,~b,~l,~p]    2            .5%             │
│ Disallowed regions          [XX]             1            .3%*            │
│                                            ----         ------            │
│ Non-glycine and non-proline residues       365          100.0%           │
│ End-residues (excl. Gly and Pro)             1                           │
│ Glycine residues                            39                           │
│ Proline residues                            30                           │
│                                            ----                          │
│ Total number of residues                   435                          │
└─────────────────────────────────────────────────────────────────────────┘
```

```
Ramachandran Plot statistics for West Nile Virus (Model)
                                        No. of residues      %-tage
Most favoured regions       [A,B,L]           309           80.9%*
Additional allowed regions  [a,b,l,p]          52           13.6%
Generously allowed regions  [~a,~b,~l,~p]      14            3.7%
Disallowed regions          [XX]                7            1.8%*
                                               ----          ------
Non-glycine and non-proline residues          382          100.0%
End-residues (excl. Gly and Pro)                1
Glycine residues                               27
Proline residues                               25
                                               ----
Total number of residues                      435
```

The results for the WNV model were 86.1% of residues in the core area of the RP, 11.3% of residues in the additional allowed areas, 2.6% in the generously area and again the number of residues in the disallowed area was eliminated (table 12). The facts that almost 90% of both models residues were in the allowed areas and that the remaining residues were in the generously allowed areas with not a single residue in the disallowed area, were evidence that the homology modelling study for Dengue and WN viruses was successful and that the two models were very accurate and suitable to be used in further modelling experiments (i.e. de novo drug design and docking experiments).

Table 12. Ramachandran Plot statistics for the improved West Nile Virus Models

```
Ramachandran Plot statistics for West Nile Virus (Model)
                                        No. of residues      %-tage
Most favoured regions       [A,B,L]           329           86.1%
Additional allowed regions  [a,b,l,p]          43           11.3%
Generously allowed regions  [~a,~b,~l,~p]      10            2.6%
Disallowed regions          [XX]                0             0%
                                               ----          ------
Non-glycine and non-proline residues          382          100.0%
End-residues (excl. Gly and Pro)                1
Glycine residues                               27
Proline residues                               25
                                               ----
Total number of residues                      435
```

The Ramachandran Plot for the Dengue Model

1A1V   Dengue Virus

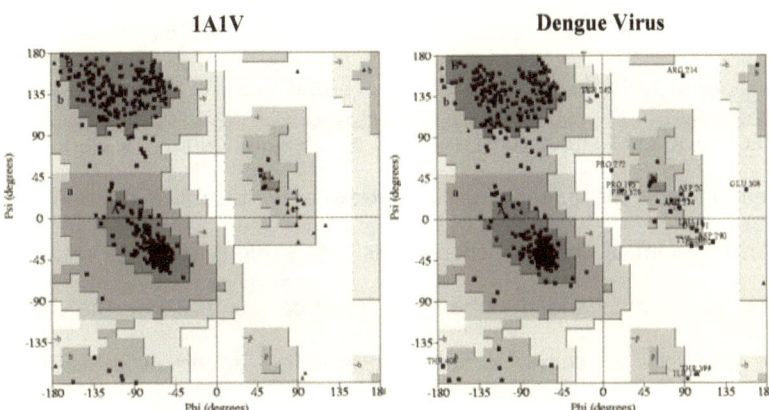

The Ramachandran Plot for the West Nile Model

8OHM   West Nile Virus

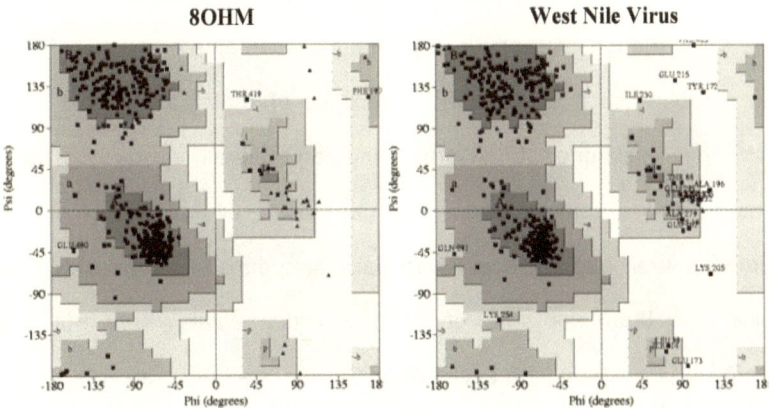

The white space on the diagram above describes the pairs or groups of atoms that have distance smaller than the sum of their van der Waals radii, because of the conformation of the backbone. As a result a sterically unreal conformation will be generated, thus making the combination of those atoms disallowed. Exception to this rule is glycine that does not have a side chain. The area with no steric clashes is represented in black color. The residues in the black area are considered to be in the allowed regions and there is where the alpha helices and beta sheet conformations are usually found. The dark grey areas include the pairs or groups of atoms with radii a fraction less than the sum of their van der Waals radii. This is the area that the motif of the left handed alpha helix would be found, where the atoms are allowed to come a little closer. The amino acids with L configuration are not supposed to form long and stable conformations although it is sometimes possible to encounter such conformation (not very often though). Usually it is glycine or asparagine or aspartate that can establish such conformation, because of their ability to H-bond with the backbone, thus stabilising an overall pretty unstable configuration of amino acids. The generously allowed areas of the Ramachandran plot are areas with pairs or groups of atoms with radii quite less than the sum of their van der Waals radii, but not significantly clashing with each other. Glycine is a quite versatile amino acid, because it doesn't have a side chain. Glycine can take phi/psi angles in every of the four quadrants in the Ramachandran plot. As a result glycines are usually encountered in loop regions in the protein, where it would be impossible for any other residue to be, because of the steric hindrance.

The model evaluation of the all of the models generated for both the Helicase and the Polymerase projects can be found in Appendix 1. Below in table 13, the Procheck report of the models before and after optimization have been summarized.

Table 13. The Procheck values of the each model after homology and after optimization and re-evaluation.

Project	Model	After Homology				After Energy Minimization & Model Optimization			
		Core	Allowed	Generous	Disallowed	Core	Allowed	Generous	Disallowed
Helicase Project	DenV	82.2	13.9	2.5	1.4	92.5	5.3	2.2	0
	WNV	80.9	13.6	3.7	1.8	86.1	11.3	2.6	0
	JEV	79.9	15.1	3.3	1.6	94.2	5.1	0.7	0
	YF	81.5	13.4	3.2	1.9	96.7	3.2	0.1	0
Polymerase Project	DenV	79.7	16.9	2.0	1.4	92.3	6.3	1.2	0
	WNV	68.2	24.6	6.1	1.0	89.8	8.4	1.8	0
	JEV	68.7	24.1	6.4	0.8	92.1	4.9	3.0	0
	YF	67.5	25.3	6.6	0.6	89.4	8.5	3.1	0

The following figures (figure 28) depict the structural motifs found on both the proteins and the models. Ideally each of the models should not be much different from their templates. The diagrams give also information on the accessibility of the residues in the proteins and finally there is also information of the PROCHECK results on the bottom of every figure. Minor differences like the ones described above are normal and are expected to exist.

Figure 28. Structural motifs, accessibility and PROCHECK summary.

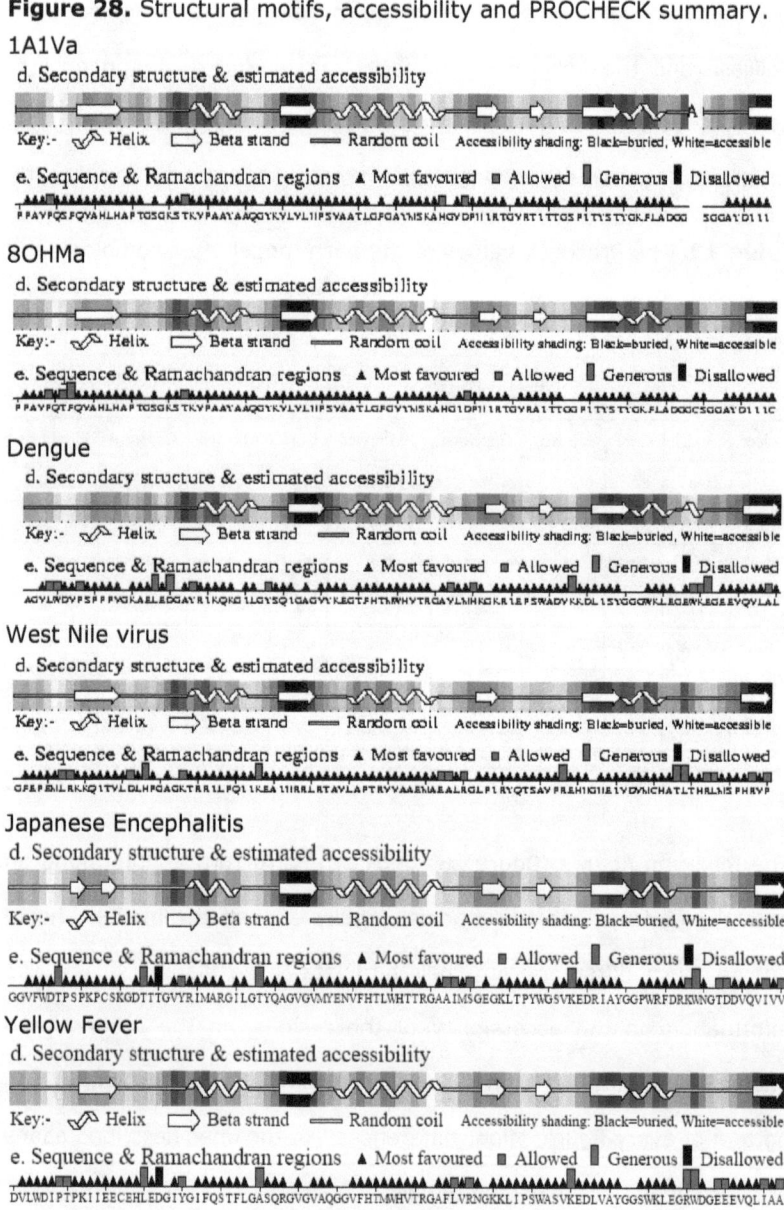

1A1Va
d. Secondary structure & estimated accessibility

Key:- Helix   Beta strand   —— Random coil   Accessibility shading: Black=buried, White=accessible

e. Sequence & Ramachandran regions ▲ Most favoured ■ Allowed ‖ Generous ‖ Disallowed
PPAVPQSFQVAHLHAPTOSGKSTKVPAAYAAQGYKVLVL1IFSVAATLGFGAYNSKAHGYDPIIIRTGVRT1TTGSFITYSTYGKFLADGG   SGGAYD111

8OHMa
d. Secondary structure & estimated accessibility

Key:- Helix   Beta strand   —— Random coil   Accessibility shading: Black=buried, White=accessible

e. Sequence & Ramachandran regions ▲ Most favoured ■ Allowed ‖ Generous ‖ Disallowed
PPAVPQTFQVAHLHAPTOSGKSTKVPAAYAAQGYKVLVL1IFSVAATLGFGVYNSKAHG1DPIIIRTGVRA1TTGGFITYSTYGKFLADGGCSGGAYD111C

Dengue
d. Secondary structure & estimated accessibility

Key:- Helix   Beta strand   —— Random coil   Accessibility shading: Black=buried, White=accessible

e. Sequence & Ramachandran regions ▲ Most favoured ■ Allowed ‖ Generous ‖ Disallowed
AGVLNDVPSFFPVGKAELEGGAYR1KQHG1LGYSQ1GAGYYKEGTFHTMWHVTRGAYLMHRGKR1EFSWADYKKDL1SYGGGWKLEGEWKEGEEVQVLAL

West Nile virus
d. Secondary structure & estimated accessibility

Key:- Helix   Beta strand   —— Random coil   Accessibility shading: Black=buried, White=accessible

e. Sequence & Ramachandran regions ▲ Most favoured ■ Allowed ‖ Generous ‖ Disallowed
GFEFDILRKKQ1TVLDLHPGAGKTRR1LPQ11KEA1IIRRLRTAVLAPTRVVAAEMAEALRGLP1RYQTSAVPREHIGIIE1VDYKICHATLTHRLMSPHRVP

Japanese Encephalitis
d. Secondary structure & estimated accessibility

Key:- Helix   Beta strand   —— Random coil   Accessibility shading: Black=buried, White=accessible

e. Sequence & Ramachandran regions ▲ Most favoured ■ Allowed ‖ Generous ‖ Disallowed
GGVFWDTPSPKPCSKGDTTTGVYRIMARGILGTYQAGVGVMYENVFHTLWHTTRGAAIMSGEGKLTPYWGSVKEDRIAYGGPWRFDRKWNGTDDVQVIVV

Yellow Fever
d. Secondary structure & estimated accessibility

Key:- Helix   Beta strand   —— Random coil   Accessibility shading: Black=buried, White=accessible

e. Sequence & Ramachandran regions ▲ Most favoured ■ Allowed ‖ Generous ‖ Disallowed
DVLWDIPTPKIIEECEHLEDGIYGIFQSTFLGASQRGVGVAQGGVFHTMMHVTRGAFLVRNGKKLIPSWASVKEDLVAYGGSWKLEGRWDGEEEVQLIAA

88

3.5 **Conclusions**

Homology modelling is not a standalone technique. There has been no homology modelling experiment with single aim to visualize a protein. All homology modelling experiments are preceding the actual main experiment. In this case the reason for doing homology experiments was to be able to do drug design experiments later on. In this case it is essential to understand exactly what is expected from a model, especially when the homology scores and sequence alignments are low. The drug design experiments that will follow homology modelling, will focus on the design of relatively small compounds (MW < 600), that will act on the RNA binding site of the helicase and polymerase proteins. So, essentially the only part of the model that is interesting in the drug design experiment is the RNA-binding site, usually referred as the active site. The accuracy in homology terms of the rest of the model is not as significant to the following drug design experiments. The accuracy of the non-active site regions of the protein were mathematically checked and corrected using molecular mechanics algorithms, to ensure that a reasonable state of quality has been achieved. The approach to designing and evaluating the active site though was significantly different. The coordinates of all the conserved and interacting residues was compared to the corresponding ones on the templates. In this case, molecular dynamics algorithms were used to explore the potential in movement of all residues and thus energetically minimize and stabilize the active site.

The Helicase project was straightforward in terms of primary sequence alignments and homology scores. On the other hand though the polymerase models did not meet the initial alignment scores and alternative routes of homology modelling had to be taken. The models were generated with MOE using the Hepatitis C polymerase structure as a template. The Hep C polymerase has been determined by crystallography at resolution 2.2 angstrom. Because of the low identity score in the sequence alignment, new techniques were used in an attempt to compensate for the low score. Firstly the ssRNA that was co - crystallized was included in the homology studies as part of the protein. The coordinates of the ssRNA were retained in the new models with the assumption that the ssRNA and each of the protein models of the same family are expected to establish the same type of interaction. The option of using the ssRNA as part of the homology study is feasible and can be considered only when dealing with proteins of the same function and role in the cell (all polymerases in this experiment) and of species within the same family (all members of the Flavi virus family). The assumption here states that when dealing with evolutionary linked species of the same family and the same protein within them, then it can be assumed that those proteins will have the same role, same function and hence similar structures and interaction patterns.

CHAPTER 4:

De De novo Drug Design

4.1   Introduction

Structure-based ligand design algorithms require huge computational resources since their operation is based on iterations and recursions. Both Iteration and Recursion run a part of an algorithm repeatedly. They are typically used when, in a part of an algorithm, something has to be done or to be calculated (i.e. the logic remains the same though the data changes) for a certain number of times. However, in Recursion, a loop is simply run for a certain number of times within the algorithm, such as printing a statement ten times is recursion. In Iteration, a loop has been specifically run, such that, the second run of the loop makes use of the computation/result from the first run, the third run (or iteration) makes use of the result from the second run and so on. Hence, in Iteration, the nth run of the loop makes use of the result from the n^{-1th} run. Genetic Algorithms are a class of computer algorithms for solving difficult search and optimization problems. The method is loosely modelled on the biological sciences of evolution and genetics and lately in the new developing area of drug design. The method is particularly useful for some of the most difficult search and optimization problems, especially problems with:

- Very large search spaces
- Discontinuous derivatives
- Multiple local optima

Genetic Algorithms (GA) work with a large population of potential solutions to the problem. The modules grow and link of LigBuilder are characteristic examples of genetic algorithms. These solutions are represented as "string" often in binary. The genetic algorithm evaluates each of these potential solutions and assigns it a fitness value. The fitness value is a measure of how well each particular solution solves the optimization problem. To form the next generation of potential solutions, individuals are statistically drawn from the existing population in a manner that favours the fittest individuals in the population. This step is analogous to Darwin's concept of Survival of the Fittest. There are many ways to perform this selection. The key is that the fittest individuals in the population have a greater chance to contribute to future generations than the least fit. The selection process is statistical though, so there is always a chance that poor performers can be selected. This diversity is critical for robust performance of the algorithm, because even the poor performers may contain critical bits of information for the optimal solution.

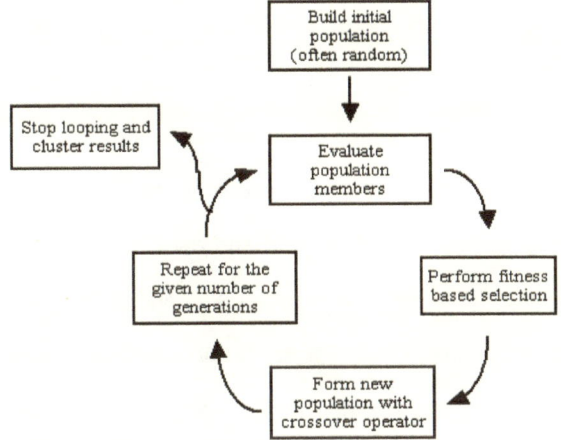

Genetic Algorithm – LigBuilder Approach

Figure 29. The steps of the genetic algorithm of LigBuilder

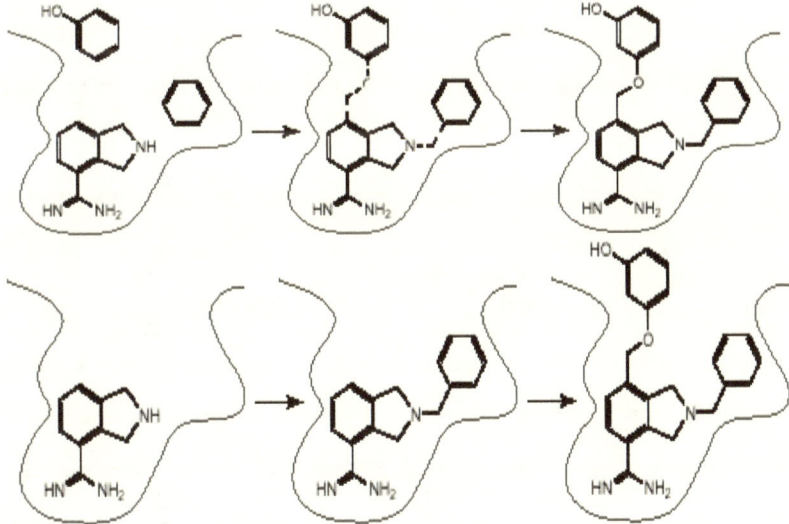

Figure 30. The two different approaches to structure-based drug design.

There are two ways to structure based drug design. The first (shown on the top) is the linking procedure, where the program selects the most suitable moieties to interact with the active site of the protein and then starts to link them together in a chemically appropriate way. The second approach is the growing procedure, which involves the determination of a group as a starting point and then the growing of a larger compound that would fit in the active site and that would be capable of establishing interactions with it. The first approach requires that the user must supply the docking/interaction points of the moieties that will be used as anchors to the final structure. The second approach requires that the user picks an efficient moiety as a starting point, and that the user directs the growing process towards the right way. The best results will be obtained from the combination of these two procedures.

Once individuals are drawn from the population, the new population is formed by combining the information from groups of selected individuals. This is done by randomly splitting the representation strings of the parents and recombining them to form new entities. This operation is called Crossover. Because the parents have been selected for their fitness, the children of these parents will, statistically, be more fit then the parents. This is done until a new population has been built out of the old one. Final step is the evaluation of the new population, which may result in starting the whole process over again by changing single or groups of parameters. The cycle described in figure 29 is continued until a solution that is good enough for the experiment's criteria has been obtained.

Many Genetic Algorithms are used primarily for search/optimization type problems. This is not really "Machine Learning" as it is commonly defined; however, the basic structure of the algorithm is the same. In LigBuilder, for example, it is possible to instruct the computer how to position the seed structure and the program will attempt to use the same approach in a different experiment. So, instead of optimization, the algorithm is used to "teach" the computer how to solve a posed problem in a relatively arbitrary environment. Many times a Genetic Algorithm in a Machine Learning environment is used for discovering effective rules for use in other systems such as heuristic and expert system type classifiers. They have also been used very effectively in the design of neural networks and in the design of training sets and regimens for neural networks.

4.2   Drug Design for the HepC Helicase – RNA binding motif

The small compound that was assumed to be a leftover of the crystallisation, gave out a very important hint for the residue that it was bound to. The small compound was attached to Cysteine431 in a very strategic position for the blocking of the passage of the RNA single strand through the helicase. The crystal of the HepC Helicase in this case was obtained with a ssDNA in the RNA channel already bound to the helicase. Still the small compound found the space it needed and was capable of binding to the Cysteine431. This means that the Cysteine431 is exposed enough to the environment of the Helicase (i.e. at least is not a buried amino acid). As mentioned before the position of the Cys431 residue was ideal, as it could be combined with the two very exposed to the solvent Arginine residues. These residues were set to define the active site of the helicase that was going to be targeted later on. The Arginine residues are capable of H-bonding - exchanging hydrogen with other compounds present in the area. The Cys431 residue was expected to establish a S-S or an S-C bond with the inhibitor. The idea was to design a compound that would interact with the two Arginine residues and the Cysteine residue, thus forming a bridge in the middle of the RNA channel in the Helicase. If the binding mode of the compound is low in energy, and all the expected H-bonds form successfully, then the compound is expected to be strong enough to block the passage of the ssRNA and as a result inhibit the function of the Helicase.

The design of the new series of molecules would result from a drug design genetic algorithm called LigBuilder (version 1.2) [50]. This program will start from a "seed" compound which can be as simple as a C atom, which must be positioned in space by the scientist though. The human factor can give rise to a variety of assumptions as well, but there is no way for the software to identify "by magic" where the active site is. A variety of different seeds was tested but with the most suitable one being the small compound attached to the Cys431. First the S-S bond between the small compound and the Cys431 was broken and the hydrogens were restored from where they were missing. The complex was minimised only for its hydrogens as the backbone of the protein was fixed. The detached compound was entered to be the starting point of the drug design algorithm and thus it was expected that this moiety would be present in this position on all the new compounds, but that was fine as it was already known that this moiety is capable of H-bonding with the Cys431. The algorithm used the small compound as a starting point and started to grow structures that would best fit the active site and establish the maximum amount of interactions with the adjacent residues of the Helicase. All the different compounds that were designed were deposited in a folder for further investigation. A similarity cut-off of 90% was used in order to make sure that structurally all the different compounds in that folder would be at least 90% different. If that was not so, there was the fear of getting a folder full of the same single compound with minor modifications each time. The compounds with the higher scores were then combined and docked in the Helicase. After a lot of repeats of the experiment a lead was obtained, which had all the moieties that could interact with the given active site on the Helicase. Even though this compound was capable of establishing so many hydrogen bonds with the Helicase it was useless as a drug. This compound had no chance to reach the target if someone would ever manage to synthesize it! This compound was used though to design a new compound that had three special characteristics. Firstly, it had a simpler structure and was reasonably synthesizable, secondly the main moieties of the initial compound that were contributing H-bonds were maintained and thirdly this new compound was designed to be diverse in terms of possible future modifications.

The standard approaches were significantly modified in order to be able to work with helicase. LigBuilder and its growing module are useless if the targeting site is not a well defined pocket. It simply keeps growing a structure that does not relate to anything. The way round this was to make use of the small molecule attached to the Cysteine 431. The fact that this small molecule is makes the Cysteine a good target. The S-S bond between the Cysteine and the small compound was broken and the hydrogens were replaced. Then the growing technique was used to generate a compound of small molecular weight. The aim was to build a bridge between the Cysteine 431 and the Arginine 393. That molecular bridge would then stand in the way of the nucleic acid, since it would be clearly blocking the way of the channel. On the other side of the channel, opposite to the Cysteine there is an Arginine (Arg393). This Arginine is an easy target in terms of being a quite capable amino acid of establishing interactions with its environment. Looking closer into the protein, it was possible to calculate the distances between the various residues that would get involved in the interaction with the ligand. Then the same procedure was repeated for the Arginine 393, where a molecule of small molecular weight was made by the growing algorithm of LigBuilder. Then the two small compounds were linked together using a linking algorithm of the same package.

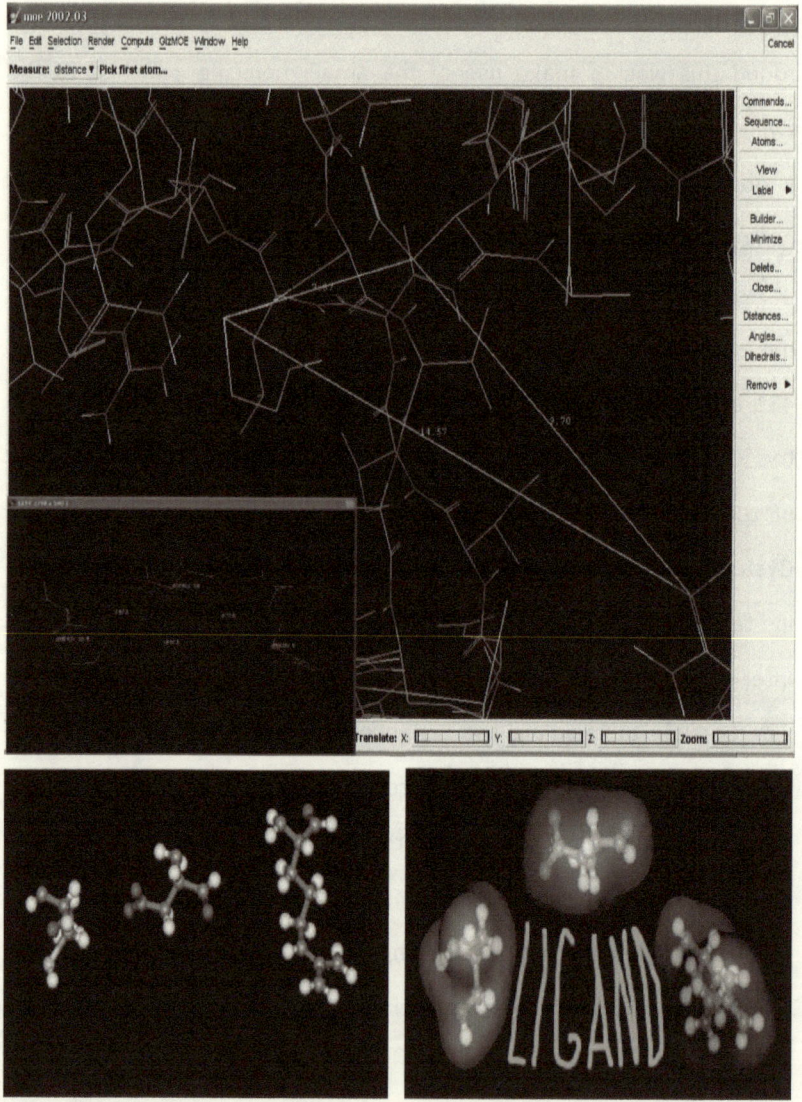

Figure 31. Distances and the available space in the Helicase's active site

The ligand should have the dimensions of that triangle and it should get stack somewhere between the Cysteine and the Arginine as it is shown in the bottom right part of figure 31. At this stage at least these amino acids had to be checked for their suitability and their reliability. The protein was checked first for its amino acid accessible surface. The amino acids that were being targeted had to be accessible enough to the solvent. Both the Cysteine and the Arginine are more than 70 percent accessible to the solvent, as it can be seen from their yellow colouring. It can be seen that the buried amino acids in the core of the protein acquire a blue colouring whereas the outer exposed amino acids tent to get a green to yellow colouring. Then the NS3 domain was checked for problems with its amino acids. The problematic amino acids are determined due to their lack of information on the PDB file. So, the software has to reconstruct these amino acids based solely on the limited information that is available for them. It would not be possible to perform any trustworthy molecular modelling experiments on reconstructed amino acids. Red-coloured amino acids are reconstructed, whereas grey ones are not. Any other colouration could mean partial reconstruction. The Arginine 393 and the Cysteine 431 are in the grey area.

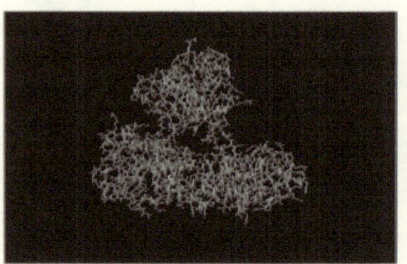 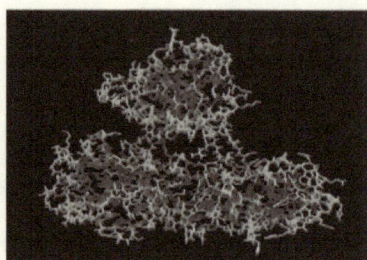

Figure 32. Helicase coloured by error-reporting colouring and accessible surface.

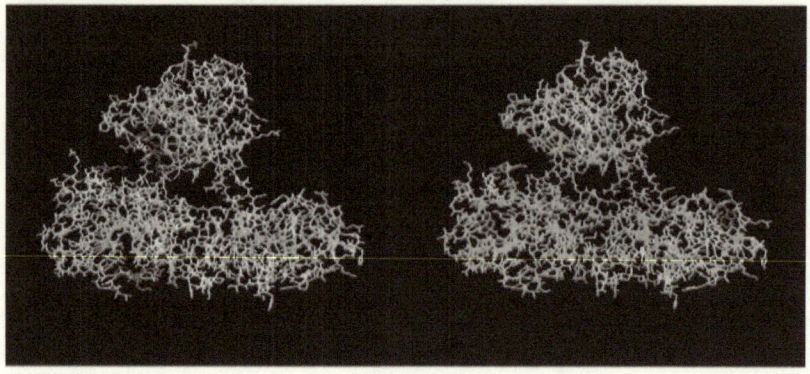

Figure 33. Helicase coloured by force-field and threading energy

The last two images of the helicase are coloured by their force-field and threading energies. Amino acids with no problems are expected to be coloured green or yellow. Again both the Cysteine and the Arginine are fine. Threading energy is a calculation of how energetically stable the amino acids are from the influences of their environment.

As soon as the two small compounds were designed for both the Cysteine and the Arginine, they were imported to the linking algorithm and the mid-part of the new compound was made. Even though this sounds like a fast procedure, it took almost 50k structures to be refined in order to get to the optimum one. A total number of about 300 generations of the genetic algorithm were used. This structure was designed, optimised and then it was docked back into the active site of the helicase. Figure 34 is the docked lead compound made with the growing and linking algorithms of LigBuilder.

The structure was made with the limited fragments located in the database available to the genetic algorithm's database. Even though growing structures for a specific target gives straightaway the advantage that the structure will pick up any available residues from the active site, and will attempt to establish an interaction with them, there is always the limitation of the fragment database. The structure was designed from a variety of approximately 60 different fragments that are available to the algorithm. If there were more fragments in the algorithm's database the structure would possibly be better and more specific for the active site of the helicase. The meaning of all these is that it is novel to use LigBuilder' s structure as a Lead Compound and still try to modify it keeping two main things in mind. First is to conserve the interactions of the lead and even to attempt to improve them, and second is to attempt to simplify it as much as possible. The ultimate purpose of all these experiments after all is to synthesise the final compounds. So, the final compounds have to be synthesisable.

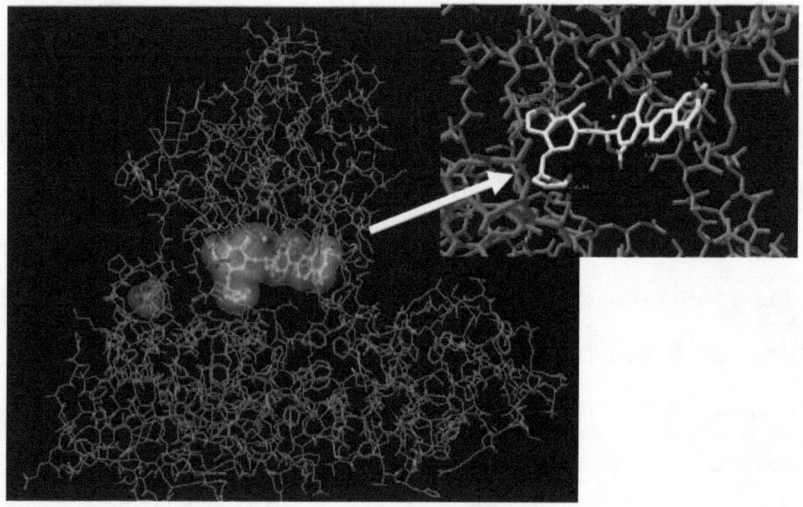

Figure 34. Lead compound from LigBuilder docked into the helicase

Below (figure 35) is the lead structure designed from LigBuilder's and the structure that derived from it after applying to it simple techniques, which mainly simplify the structure, thus making the compound synthesisable, while trying not to lose the all important moieties on the lead compound structure, which would mean loss of interaction with the helicase's residues and consequently loss of the compound's binding affinity.

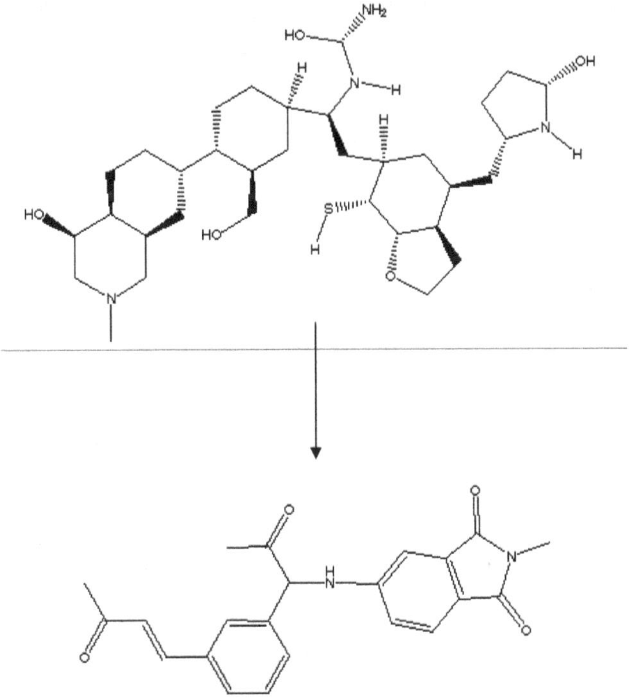

Figure 35. The simplification and refinement of the lead compound from LigBuilder. Keeping the interactions, while making the molecule synthesisable.

The last part of this compound-designing procedure was to use even more simplified structures that could still dock in the active site of the helicase. Designing the easiest to synthesise compound, which can still dock in the active site, even with reduced binding affinity, is always a very useful reference. The compound that was designed is shown in the figure below. This compound had to be made, so that it could be used as a reference structure in the future.

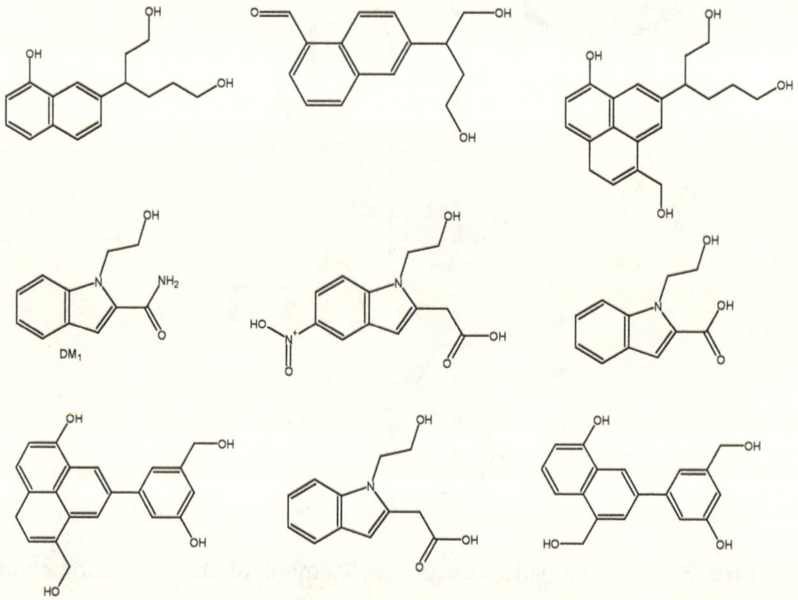

Figure 36. List of compounds designed with LigBuilder for site1 on the Helicase

4.3　　　Drug Design for the HepC Helicase – Application of the "tube"

The application of the tube in the channel of the helicase significantly simplified the job of the Growing genetic algorithm of LigBuilder. The problem of the compound growing towards all directions was eliminated. The only parameter that had to be investigated was whether or not the tube would affect the GA. The tube was originally made from carbon atoms allocated in space in order to make the tube. The PDB file of the tube was opened with a text editor and the atom characteristic from C (Carbon atom) was changed to Du (Imaginary atom). Du atoms do not have any of the standard atomic properties. They are incapable of establishing interactions of any nature and they have no charge. Their only function is to define the space that is available for the genetic algorithm to work in. The original purpose of the "Du" atom was to take the place of any atom that lacks information on a PDB structure, thus making it unrecognisable by the various visualisation programs. The results of the application of the tube were astonishing. The structures that LigBuilder was generating were a lot simpler and more drug-like. Finally a very simple and rigid compound was obtained. This compound was docked back into the protein with the tube absent this time, and it proved to be the most stable docking achieved for the helicase so far. The original structure of this rather simple compound was then enhanced with an extra carbon on the PARA position of one of the two aromatic rings. This carbon was supposed to get involved in an interaction with the accessible Cysteine 436 residue. The compound with the extra carbon was docked into the helicase and it proved to be the second most stable bridging compound between the Arginine and the inner part of the helicase. The stability of these compounds derives from their simplicity and rigidity. The arrangement of the double bonds on these compounds leaves very narrow space for rotation of any part of the compounds. The degrees of freedom are very few. Figures 44-45-46 show the two versions of this compound with and without the extra carbon. It is obvious that the presence of the extra carbon has pushed the compound a bit lower, which is evidence that the extra carbon successfully established an interaction with the nearby available sulphur from the Cysteine residue.

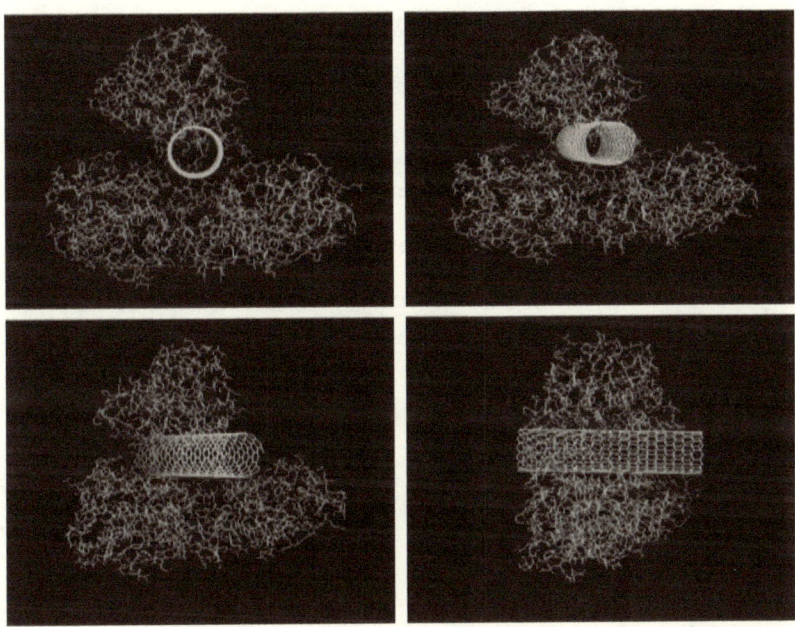

Figure 37. The Application of the tube simplified the task of LigBuilder.

4.4   Drug Design for the Dengue Helicase

The drug design process was done with the structure of the known class of inhibitors reported by Virofarma (see 5.2.2). A combination of structure-based and lead optimization drug design techniques were used. Firstly the lead had its chemical structure modified to either slightly or significantly altered ones, then those compounds were screened *in silico* and docked on the helicase. From Appendix 8 it appears that the dengue helicase has the same behavior with the HepC helicase. Smaller compounds than the lead and compounds with rigid linker are not favored.

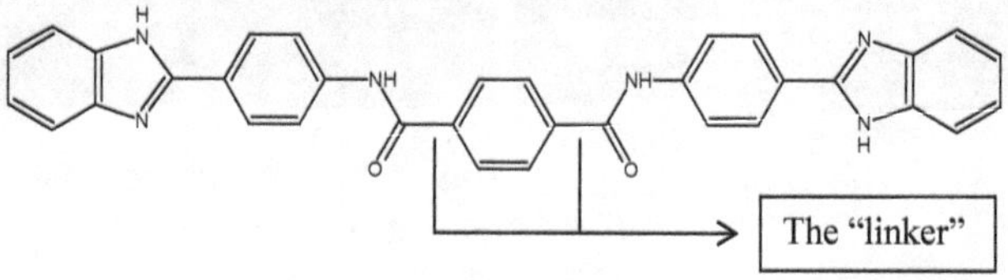

Figure 38. The lead compound with designated the variable linking part of it.

So, the favored compounds were the ones that had big and long carbon chains as linkers. That can be explained by taking into consideration the size of the helicase, where long and flexible compounds could compete with RNA binding. Moreover it was found that a compound with one or both the imidazole rings substituted with carboxylic group is capable of interacting with the receptor, whilst having a tight conformation binding as well.

Besides the ligand-based approach to drug design for the Dengue helicase, structure-based drug design experiments were also performed on the Dengue Homology model. The interaction pattern between the helicase and the ssRNA were mapped with the aid of a contact interaction tool within the MOE suite. All the interacting residues with the ssRNA were isolated and separated in a new file. The inner region of the active site was selected and a multi-fragment search was performed using all entries in the database (counting 202 moieties - MOE 2004.03). The fragments were scored and the best

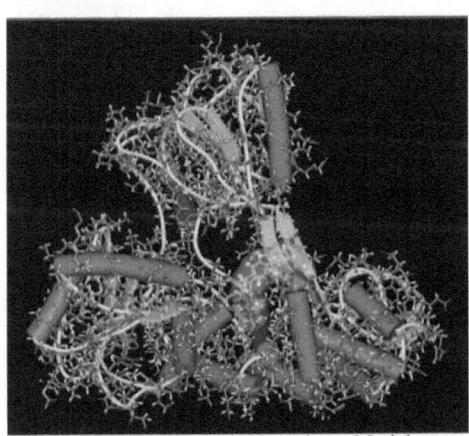

ones were selected to be included in a drug design experiment using a linking algorithm of ligbuilder. Four different moieties placed by the multi-fragment search were chosen to be used in the drug design process. The complexity of the structure was forced to be low, by choosing small

Figure 39. The Dengue Homology Model was used for structure-based de novo drug design. The final lead after optimization is shown in transparent surface representation, between domains 2 & 3.

molecular weights and by limiting the number of interaction between the future compound and the protein to those found by the multi-fragment search. The number of H bond donors and acceptors was estimated by the Lipinski's rule of five initially and later on by limiting the number of H bonds to the ones already acquired by the multi-fragment search. This made the drug design approach using the linking algorithm very fast and ensured that all new suggested compounds will be simple in structure and thus synthesizable. After the first compound was obtained, it was energetically minimized within the fixed and charged active site of the helicase. First minimization was performed with a molecular mechanics algorithm (MM+, hyperchem suite) and then with a semi-empirical algorithm (PM3, hyperchem suite). Then another drug design algorithm called grow (ligbuilder suite) was used, to increase the complexity of the lead compound and to look for new possible interactions. All the potential H bond donors or acceptors within range were highlighted and efforts were made to take advantage of them by adding new moieties on the most suitable location of the lead compound. After repeating the same process for a numerous times a new improved compound was obtained. The efficiency of interaction of that new compound with the helicase 's active site was determined by docking the compound with MOE. It was finally observed that every H donor or acceptor on the final compound was involved in H bonding with the receptor, while still being synthesisable.

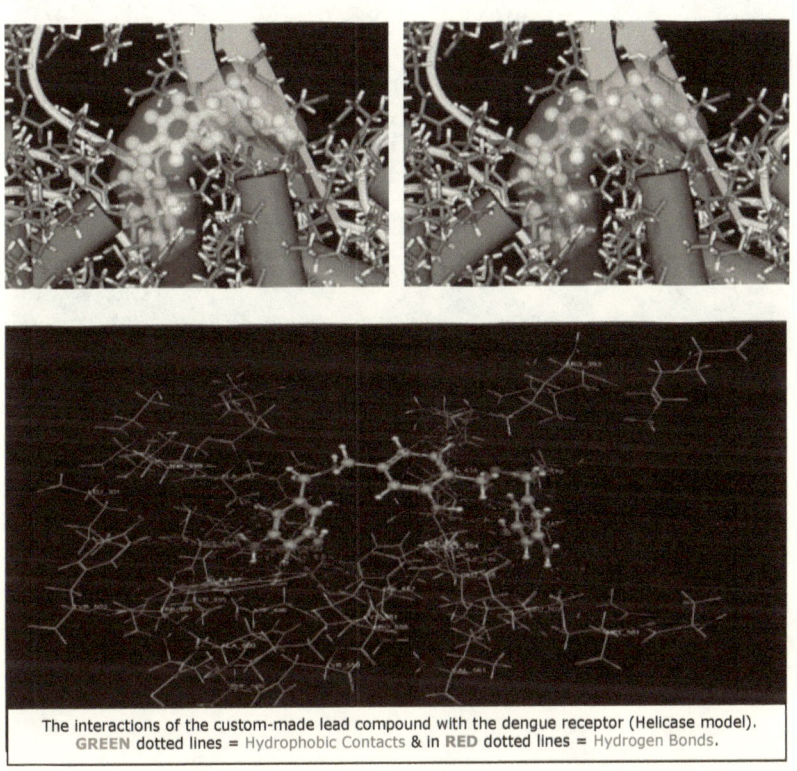

The interactions of the custom-made lead compound with the dengue receptor (Helicase model).
GREEN dotted lines = Hydrophobic Contacts & in RED dotted lines = Hydrogen Bonds.

Figure 40. The Lead Compound designed for the Dengue Helicase Model. This Compound was designed using the structure based drug design approach using LigBuilder. It is remarkable that every atom that is capable of hydrogen bonding is involved in one. The synthesis of this compound is currently on the way.

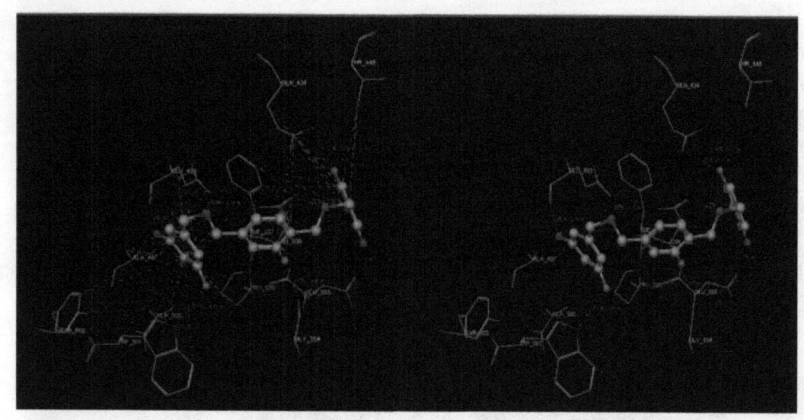

CHAPTER 5:

Molecular Docking & Screening

5.1   Introduction

There are many approaches into docking small drug-like compounds in protein active sites. They are mainly divided in three groups, the rigid docking, the semi-flexible docking and the flexible docking. Rigid docking is when neither the conformation of the ligand nor the active sites is altered. This approach is considered to be fast but not accurate. Semi-flexible docking involves giving full degrees of freedom to the ligand but keeping the residues of the active site unaltered. This approach is more accurate than the first one, takes more time and computational effort but makes a huge assumption, which says that the conformation of the active site does not affect the conformation of the ligand. The final conformation of the ligand in a docking experiment depends on the co-evolution and co-optimization of both the ligand and the protein's residue. The latter describes the concept of flexible docking. It is a co-minimization of the energies both in the ligand and the protein molecules. The aim is to reduce the energy of the system as much as possible by changing different conformations for both the ligand and the protein, so that a low-energy, stable state has been reached for the ligand-protein complex. Flexible docking is done with the aid of genetic algorithms, as described above. An attempt to match positions of establishing possible interactions between the ligand and the protein, while at the same time using molecular dynamics simulations, trying to keep the complex's energy as low as possible.

Predicting how a particular compound will interact with a certain protein is a laborious and time-consuming task, since there are endless parameters to be taken into consideration. A newly developed approach to speed up drug discovery that has recently been introduced is the High Throuput Screening (HTS) in the ChemInfrormatics field. HTS and ChemInfrormatics work on the hit-n-miss approach of fast and rigid docking of thousands of compounds in a non-flexible protein active site.

Chem-informatics is a new but rapidly expanding field in drug design. It is Driven by technological and method advances, *e.g. combinatorial chemistry* (CombiChem) and *automated synthesis*. Regularly deal with 10^4-10^6 molecules in one project, rather than 10-100s considered normal just ten years ago. Generation of such large numbers of compounds in such small amounts also requires new testing methods High-throughput screening (HTS) is a generic term for rapidly testing the activity of CombiChem results. HTS is now the primary means by which pharma- & agro-chemically active compounds are discovered. HTS is any screening method for >1000 compounds per day. The first step is that the Disease/enzyme target is identified and specific assay(s) are determined.

5.2   Hepatitis C

The target for this study was the x-ray structure of the HepC Helicase with the code 1A1V. The NCI Library of all the anti-cancer compounds synthesized until the first months of 2003 were tested for high activity. The database contained entries of all the tested compounds against cancer, not only the active ones. The NCI database of anti-cancer compounds that was used was in SMILES format. The first task would be to convert the smiles database in a pdb-like compound database. The most appropriate format was chosen to be the Tripos [53] *.mol2 format for its versatility and enhanced data storage pattern.   The conversion was done using a MOE script on HELIX. The smiles database was converted in a mol2 database of structures.

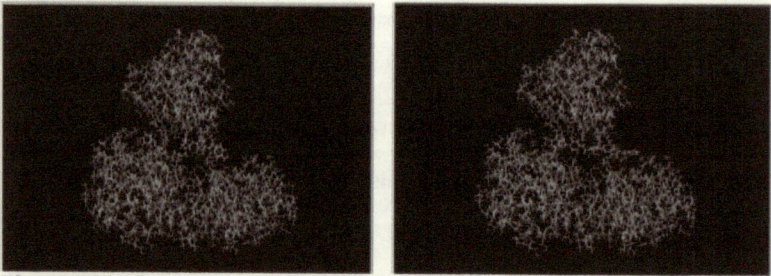

Figure 41. Two of the lead Virofarma Compounds docked in the HepC Helicase (see table 14 of compounds)

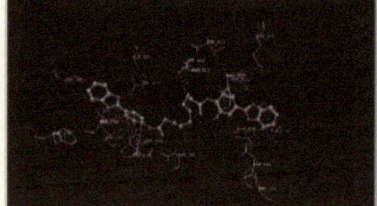

The number of compounds in the NCI database is approx. 265,000 compounds. An additional database of about 1,000 extra anti-cancer compounds was also screened. This latter was also obtained from NCI. It contained compounds in a mol2 file format, but in all saved in a single file, which had to be split using a simple awk script. So, the final number of compounds available for HTS was 266,000. The docking and energy-calculation steps would be performed either by MOE or FlexX in Sybyl. Below, table 14, summarizes the results from docking the same 7 lead compounds in the HepC Helicase using different algorithms. Successful docking is assumed to be the ones with correct ligand conformation & interaction pattern with the receptor.

Table 14. The 7 lead compounds docked with different algorithms.

STRUCTURE	LINKER	FlexX	FlexiDock	MOE	GRAMM	BiGGER[†]
	-benzene-	√	X	√	√	X
	-butene-	√	X	√	√	X
	-C2-	√	√[‡]	√	√[‡‡]	X
	-C4-	√	√[‡]	√	X	X
	-C6-	√	√	√	X	X
	-C7-	√	√	√	X	X
	-C8-	√	√	√	X	X

[†] part of the Chemera suite. [‡ & ‡‡] optimum conformation observed between 30-60[‡] & 60-90[‡‡] docked conformations

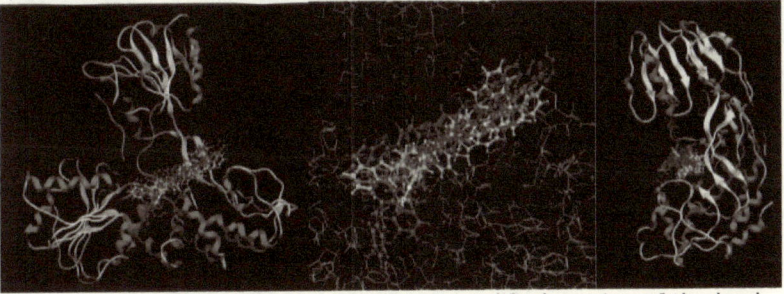

Figure 42. The figure on the left is the simplified version of the lead and on the right there is a modification of it.

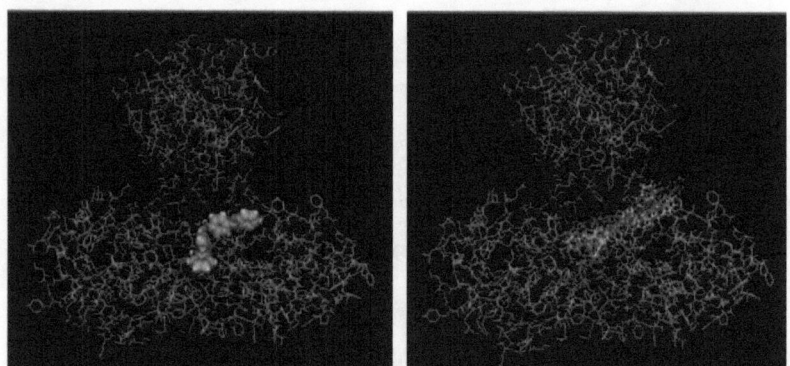

Figure 43. The figure on the left is the simplified version of the lead and on the right there is a modification of it.

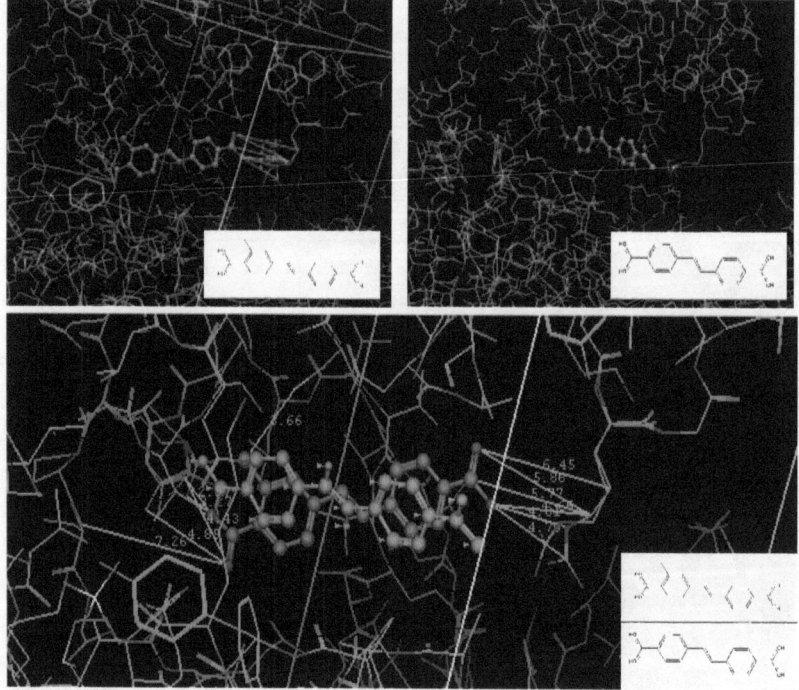

Figures 44-45-46. Upper Left : The tube-derived compound with the extra C.
Upper Left : The original tube-derived compound
Bottom : The two above compounds superimposed

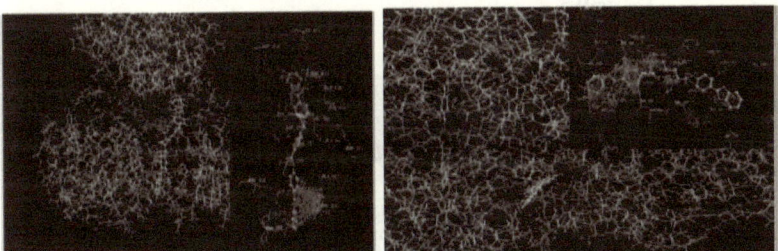

Figure 47. One of The Virofarma Compounds docked in the HepC Helicase

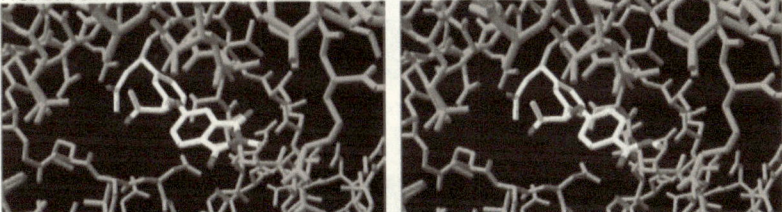

Figure 48. The figure on the left is the simplified version of the lead and on the right there is a modification of it.

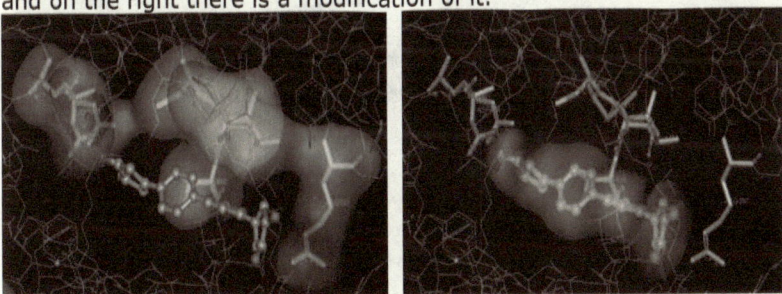

Figure 49. The docking the symmetric compound designed for helicase

Figure 50. One of the above compounds, docked into the Helicase's site 1.

5.3   Dengue

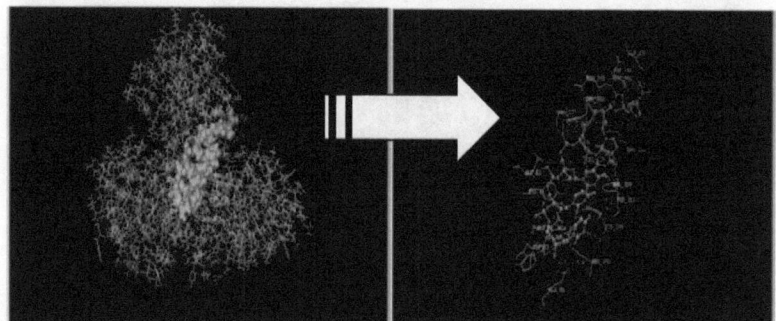

Figure 51. The Virofarma Compounds X & X docked in the HepC Helicase

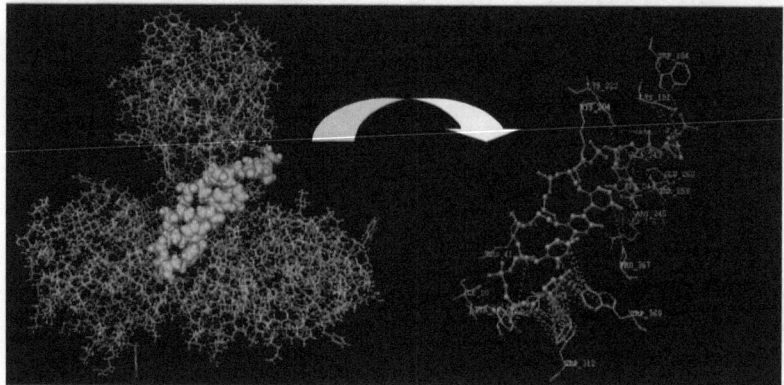

Figure 52. The Virofarma Compounds X & X docked in the HepC Helicase

The single strand of DNA was removed from the HepC helicase and was docked into the Dengue Helicase model, in an attempt to judge the quality of the model. The ssDNA was successfully docked using simulated annealing in standard conditions and the interaction was confirmed with a molecular dynamics simulation that the ssRNA endured and stayed in place, not losing any interaction. Then the interactions of the ssDNA with the protein were calculated and it was proven that they were very similar to those of the HepC. The same work was repeated with the West Nile virus model as well and the results showed that the ssDNA was not be able to pickup some specific interactions from the West Nile virus model, as it did with the HepC and the Dengue. This can be explained from the fact that the West nile virus has little homology with the HepC and the Dengue in the RNA binding domain. Still the most important residues and hydrogen bonding is present in the ssDNA – West Nile model interaction.

Then the lead compound was docked in both the dengue and the West Nile virus models and it showed the exact same binding mode with almost the exact same interaction pattern. After the screening *in silico* of the 51 modified compounds (in Appendix 8), it was found that basically all helicases including the HepC, the Dengue and the West Nile virus one have the same preferences in ligand binding. There are slight differences in the binding affinities on the 51 modified compounds with the 3 helicases. Especially with the case of WNV, the reduced affinity for binding of certain compounds may be due to limitations in the model of the WNV helicase.

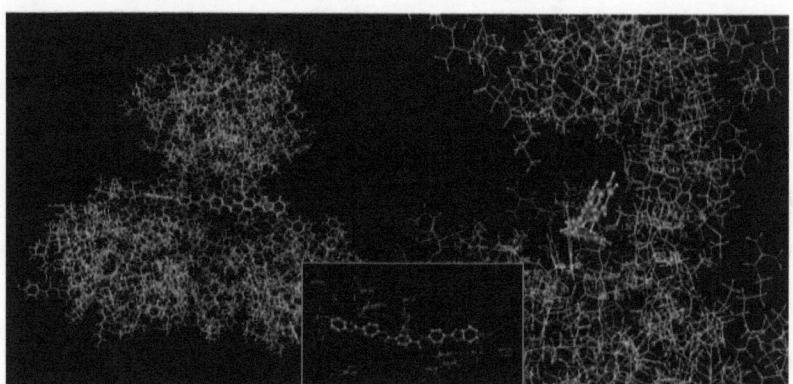

Figure 53. The Virofarma Compounds docked in the Dengue Helicase

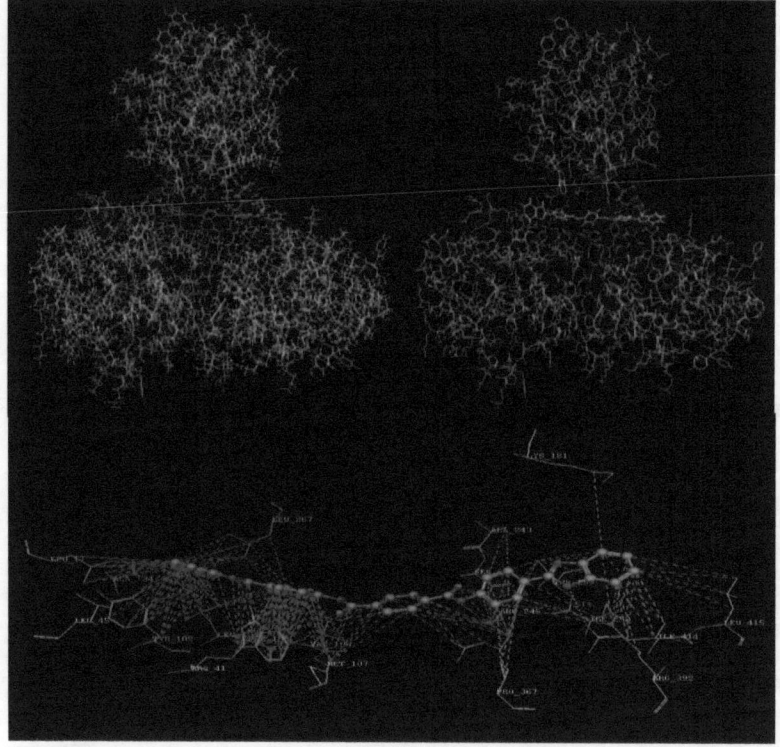

Figure 54. The Virofarma Compounds docked in the WNV Helicase

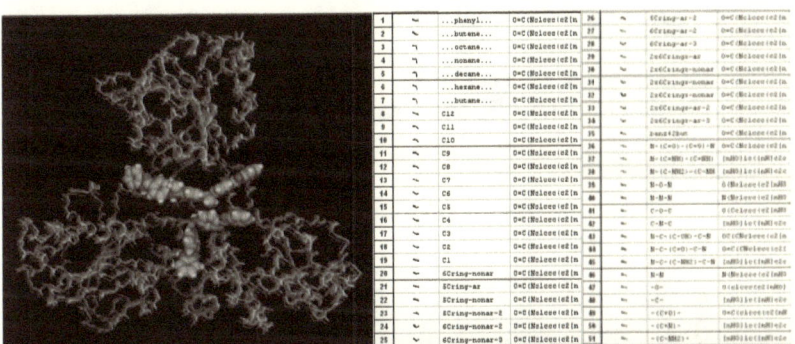

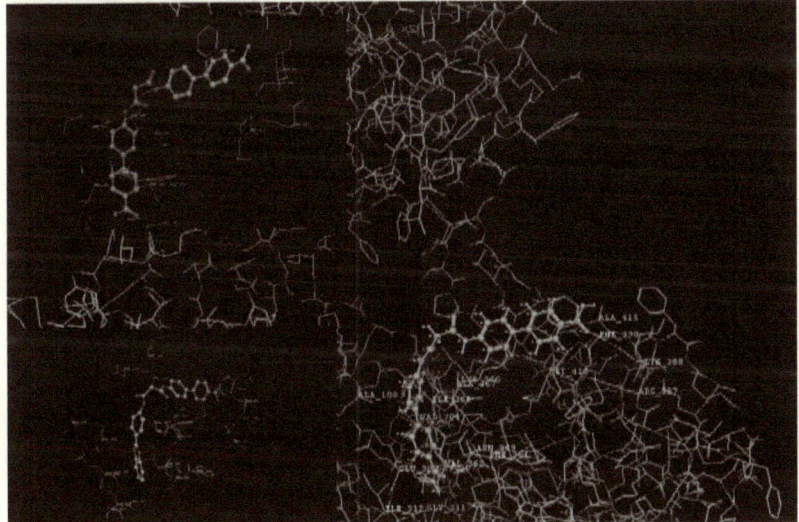

Figure 55-58. Two of the 51 modification of the lead compounds docked in the Dengue Helicase model. In the right hand side it is shown how it fits on the Dengue Helicase and on the left hand side they interactions that they pickup.

Chapter 6 has been edited by
Dr Chrisanthy P. Vlachaki

CHAPTER 6:

6.1 Introduction

The genes of the Dengue type 2 NS3 domains (full – 1,8kb) and Helicase protein (separate – 1,3kb) were obtained from a cDNA library, from the Pasteur Institute in Paris, in order to produce the recombinant proteins that these genes code for. The cDNA product was amplified by PCR and ligated into the pET23-b plasmid and later on the pGEM-T plasmid with built-in ampicillin resistance. The plasmid was transformed into E.Coli bacteria (DH5-α) strain and the bacteria were plated out in ampicillin containing plates. The only surviving bacteria the next day should be those carrying the pGEM-T plasmid. Single, healthy-looking colonies were picked up and cultured overnight in liquid suspensions (cloning). Then the cells were lysed and their plasmid was extracted (mini-prep). The molecular weight of the plasmids was confirmed by gel electrophoresis. The DNA was then purified and concentrated to be sequenced. In order for the sequencing reaction to work, special DNA primers have to be designed. Here for the two genes of the NS3 domain and that of the Helicase, four different primers are being used. Three of these primers were sequence specific and came with the genes, but one of them had to be designed especially for the needs of the experiment. The gene specific primers are the NS3F the NS3R and the HelF. The fourth primer was custom made and was named HelIntF. Sequencing checks are essential in order to eliminate the possibility for a mutation from the DNA carried on to the protein. A mutation at the DNA level would result in at least a wrong amino acid in the coding protein, resulting in a mutant protein, or in the worst case scenario in the coding of a stop codon that would stop protein synthesis completely.

So, the main steps to the Preparation on the Dengue Helicase and the Dengue NS3 domain gene cloning were the following:

1.) Design the primers to amplify the gene by PCR (In this case universal M13F and M13R primers were used; those primers will anneal on the plasmid just before the gene ligation sites).

2.) Isolate enough genomic DNA for the PCR work

3.) Amplify the gene by PCR

4.) Purify the PCR product from the primers

5.) Insert the PCR product into an appropriate vector for expression (the vector chosen for this project was the pGEM-T)

6.) Transform bacteria (usually *E. coli*) with the pGEM-T + gene vector.

7.) Purify the vector again from the bacteria. Check the clone by sequencing to confirm that in the ligated genes no mutations have been introduced.

7.) Expression of the gene product

8.) Purification of the gene product.

8.) Enzymatic assay ready to accept compounds for testing.

6.2 Polymerase Chain Reaction towards Gene Amplification

The genes of the Helicase protein and the NS3 full domain had to be amplified to be used for the ligation with the plasmid. In order to amplify the two genes a PCR experiment was setup. The polymerase chain reaction involves many rounds of DNA synthesis. So, in the end of the experiment there will be much more DNA available. For the DNA synthesis reactions the two genes were used as templates. Universal M13F and M13R primers were used and taq polymerase. The two primers annealed at each end on the NS3 & the Helicase genes and the PCR reaction went on for 30 cycles. The cycle used was the default one involving the denaturing of DNA at 95 °C for one minute then cool down at 58 °C for one minute in order to initiate primer annealing and then heat at 72 °C for two minutes to allow the DNA polymerase to extent the DNA.

6.3   Vector Preparation I – Plasmid Restriction & Phosphatase Reaction

Restriction enzymes are the main tools used for cloning.   These enzymes are able to cut DNA at specific sites - like a pair of scissors. Once a gene (or usually a smaller piece of a gene) has been cut, it can then be pasted (using an enzyme called DNA ligase) into a special piece of artificial DNA called a plasmid.   In this case the plasmid of pET23-b was cut with HindIII. pET2 has only one restriction site for this restriction enzyme. As a result after the restriction reaction the plasmid should be linear. To stop the plasmid for getting back together to its original circular shape a phosphatase reaction was performed straight after digestion. The phosphatase reaction removed the Phosphate groups from the 3' ends of the restricted plasmid so

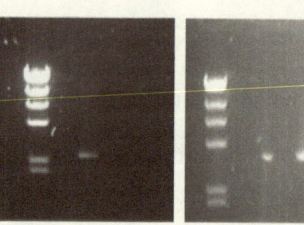

Figure 59. The digested vector by Gel- Electrophoresis

that the to lagging ends could not get back together. The insert has got phosphate groups at both of its ends and this way when the insert was applied there was no problem for it to ligate to the plasmid.

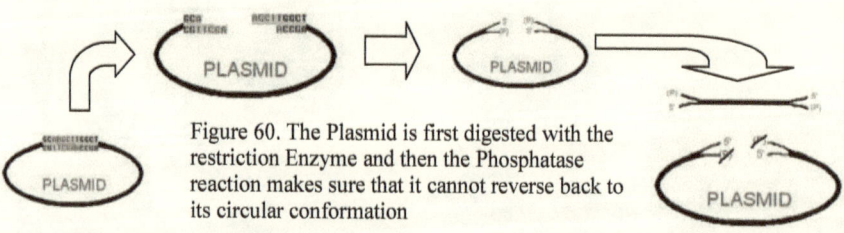

Figure 60. The Plasmid is first digested with the restriction Enzyme and then the Phosphatase reaction makes sure that it cannot reverse back to its circular conformation

132

6.4   Vector Preparation II – Gene Insertion

The process of gene insertion is done with the aid of DNA ligase. The genes of the helicase and the NS3 domain had to be incorporated to a plasmid in order to initiate the bacterial cloning process. The process of joining DNA fragments together with covalent bonds is known as ligation (figure 61). In a ligation reaction a phosphodiester bond is build between the 3' hydroxyl end of one nucleotide and the 5' phosphate end of the other nucleotide. The enzyme used in the

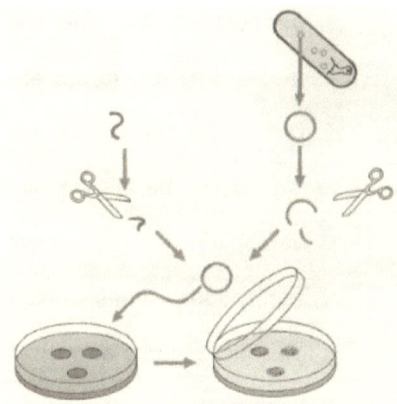

experiment was the T4 DNA ligase from the T4 bacteriophage. This enzyme anneals fragments with overhanging cohesive ends. The plasmid (which now contains the gene) is then inserted into a bacterium. As the bacteria grow and multiply, they make many copies of the plasmid (and hence

Figure 61. Gene ligation in the Plasmid

they also make many copies - clones - of the gene). In this way, whole genes, or small parts of genes can be cloned for other work like gene sequencing or protein production (since genes carry the sequence for making proteins). 2-hr and Overnight Ligation methods were used and were found to be as efficient (see experimental).

6.5   Transformation

Bacterial Transformation is the process where competent bacteria take on DNA (usually in the form of plasmids) from their environment. However DNA may get into the cells from different sources other than plasmid transformation. For example, bacteria can take over DNA debris from lysed (dead) cells in a bacterial colony. The term competent means that bacteria have been treated and their cell membranes are more permeable to genetic material than before. Here the E.Coli strain DH5-α was used, because of its versatility, short replication times, high plasmid content and most importantly for its compatibility with the chosen plasmids. The

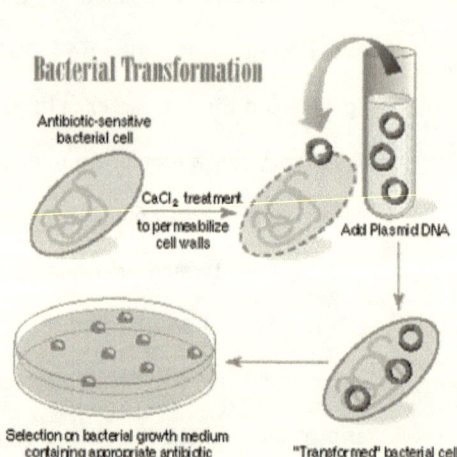

Figure 62. Bacterial Transformation

E.Coli cells were initially transformed with both genes on the pET23-b plasmid (each on different plasmid). The overnight culture was setup, but it seemed that the number of colonies on the AMP plate were limited. The same procedure was repeated with the pGEM-T vector, which apparently gave overnight cultures with colonies TNTC (too numerous to count). The transformation was done by the "heat shock" method. According to this method the plasmid is added to frozen (-72 °C) cells and they are kept on ice for 45 minutes, and then the cells are placed in a 37 °C waterbath to permealise their membranes. After a short incubation the cells are inoculated on a petri dish with LB and AMP antibiotic.

6.6   Microbiology

There are two types of cell culture involved in this experiment series. The first is the inoculation of transformed bacteria on LB agar petri dishes, in order to generate overnight bacteria colonies that will carry the inserted gene. To ensure that the bacteria have been successfully transformed with the vector, ampicillin antibiotic is also added to the LB agar medium. E.Coli cells of the DH5-α strain do not have natural resistance to Ampicillin antibiotic. The pGEM-T and the pET23-c vectors though have the AMP resistance gene built-in their sequences. As a result only the cells that have uptaken the plasmid (with the helicase or the NS3 gene) will survive. The second type of bacteria culture system is the liquid suspension system, where single bacteria colonies are picked up from the overnight inoculation on LB agar with AMP and are suspended in liquid LB containing AMP as well. On both occasions it is vital that the timing of the cultures is accurate, because the competent DH5-α E.Coli cells have very small and reduced life cycles. This favors the quick replication of the inserts along with the cells, but makes the balance of the cell culture systems very fragile. Stopping a culture too soon will yield only few DNA, since the number of cell replicating per unit time are more time by time. But there is the problem of utilizing all the available resources of the media and reaching lag phase in very fast.   The optimized culture time for both type of cultures with the given genes, the given bacteria strain under the standard condition was found to be 13 to 14 hours from inoculation or suspension.

6.7 DNA Extraction and Cell Lysis (mini-prep)

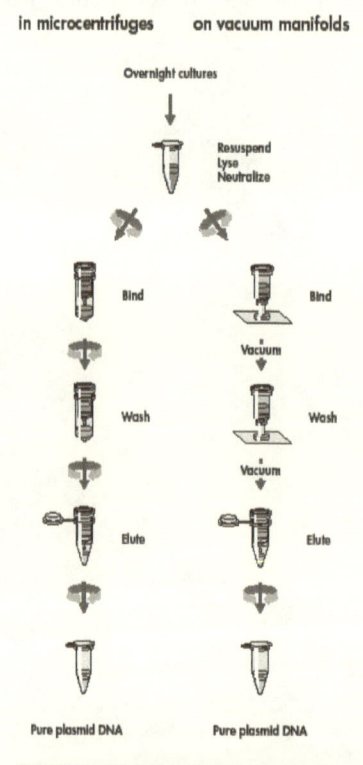

Figure 63. The mini-prep steps of DNA extraction

After the overnight liquid suspensions millions of cells have been created, each one of them carrying multiple copies of the insert gene. The extraction of the DNA plasmid from the bacteria cells was done using the mini-prep method. The bacteria are lysed and their cell contents are released in a cell – organelle suspension. Cell debris is removed by centrifugation. The supernatant liquid (containing the plasmids) is passed through a column and the DNA get trapped in it. Adding a series of reagents (see experimental) it is possible to wash firstly DNA from any alien leftover materials from the cells and finally to wash it off the column and retrieve it purified. The mini-prep of the pET23-b vector showed that this vector is a low copy one for the specific E.Coli strain used in this experiment. So, for the needs of molecular biology experiment, it was decided to use the high copy pGEM-T vector. The pET23-b vector will be used again in the Biochemical part of the experiment during protein expression, since this vector is considered to be more efficient for protein expression that the rather generic pGEM-T vector. The gel electrophoresis picture below is the picture of the mini-prep products. It is clear that both 1,8 kb NS3 domain genes and 1,3 kb helicase genes are present on the gel.

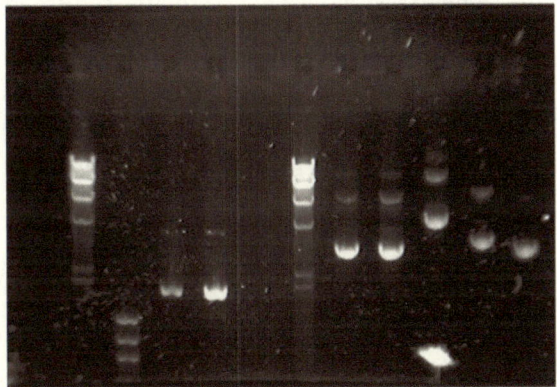

Figure 64. The 1,3 and 1,8 genes on Gel Electrophoresis

6.8   DNA purification and Concentration

For the purification and cleaning of the isolated DNA a DNA precipitant commercially available was used (see experimental). The DNA precipitant is a combination of Phenol Chloroform and Isoamyl Alcohol. After the addition of Phenol Chloroform the DNA solution was centrifuged at 13000 rpm for 10 minutes. All remaining cell debris should stay at the bottom of the eppendorf tube and carefully removal of supernatant will ensure that the DNA is pure.

For the Concentration of the DNA a combination of Isopropanol and NaAccatate was used. DNA is left in the fridge with the above reagents to help it precipitate faster and then it is spinned at 13000 at a benchtop mini-centrifuge for 20 minutes. Then 70% ethanol is added and the solution is spinned for another 5 minutes. The eppendorf is heated in a PCR block to evaporate ethanol at 52 °C.

6.9   Cloning

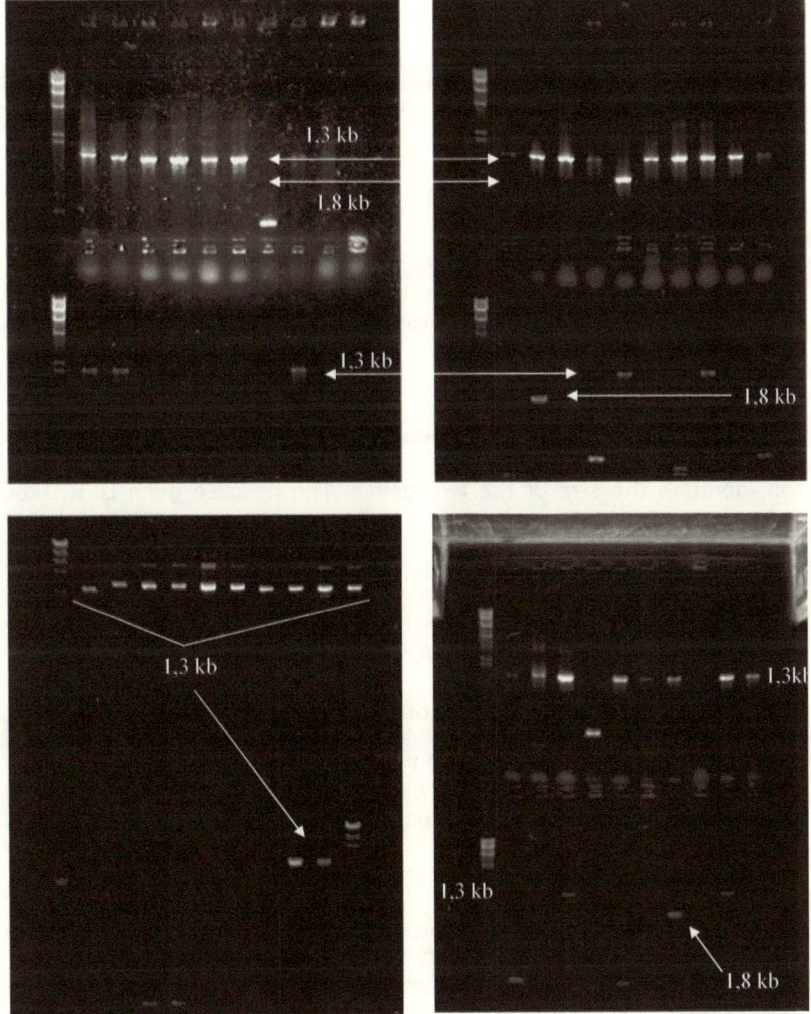

Figure 65. Cloning of bacteria with the genes of the NS3 domain and the Helicase Proteins. Each of the bands on the gels accounts for a different clone.

The cloning of the two genes for the Helicase protein and the NS3 domain was done using the same host (E.Coli DH5-alpha strain), the same vector (pET23-b initially and pGEM-T later on), under the exact same conditions. Still it appears from the yields of the colonies that the bacteria transformed with the helicase gene are much more efficient in growing and reproducing copies of themselves (clones). Even though the cultures themselves seem to be healthy in both cases (same dark yellowish color and viscosity) the NS3 cells are struggling to grow.

An explanation to this phenomenon can be given by taking into consideration the size of the two genes. The helicase gene is 1,3 kb long whereas the NS3 domain gene is 1,8 kb long. Since all the rest of the parameters are not changed it appears that this difference of 500 bases in DNA length makes a different in the efficiency of the cloning of the bacteria cultures. The size of the pGEM-T plasmid is about 4 kb long. So, this difference of 500 bases is quite significant when compared to the size of the plasmid. What is not clear is whether the increased size of the gene causes trouble to the plasmid or to the cells during transformation. There is the option that the cells are competent enough to take up to a certain size of foreign plasmid inside them. It may be the case that adding the 1,3 kb gene onto the 4 kb plasmid is well within range of the competency of the cells, whereas the addition of the slightly larger 1,8 kb gene is marginal to the ability of the cells in uptaking foreign DNA.

Another crucial discovery of the cloning process is the suitability of the two plasmids used in this experiment. It was found that the pET23-b plasmid is much less efficient than the pGEM-T plasmid under the exact same conditions (same genes, cells and procedures). The reason is unclear but it may be that the cell transformation protocol that was used for cloning could favor one plasmid over the other.

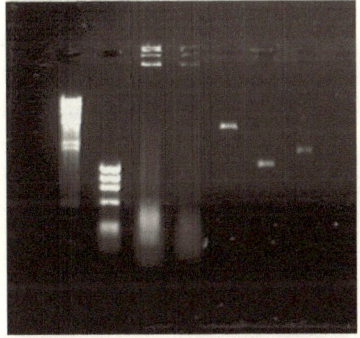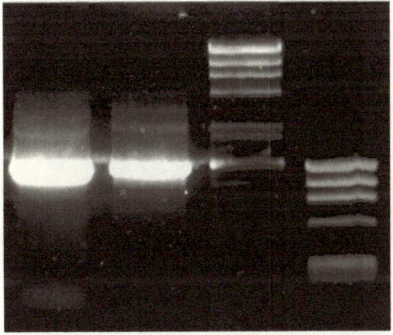

Figure 66. Top Band: the Vector, Bottom Band: The 1,3 gene and Middle Band the 1,8 gene.

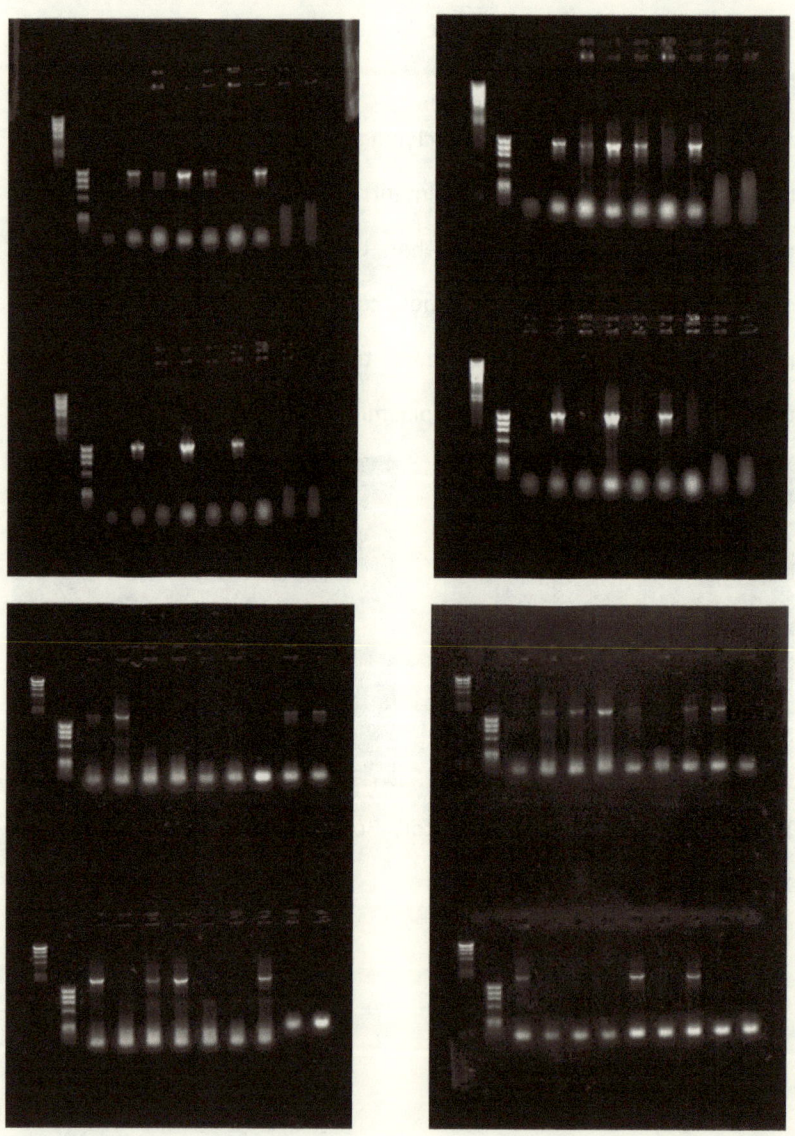

Figure 67. The PCR screen to select gene (PCR) positive colonies. (all bands above are of the 1,3 kb gene)

6.10 Primer Design I

The sequencing reactions are only accurate for 400-500 bases one way. So for the case of the helicase protein the two available primers (the HelF and the NS3R) would not be able to meet in the middle of the gene, since the helicase gene is 1,3 kb. Things are the same for the NS3 domain that is 1,8 kb long. Both genes will have a "blind region" for the sequencing reaction in the middle of the helicase domain. The protease of the NS3 domain will be covered by the NS3F primer. As a result a primer had to be designed for the middle of the helicase domain. The steps in primer design are summarised below:

— Select a region in the area of interest

— Apply the formula:

— $Tm = [4 (G + C) + 2 (A + T)]$

— Make sure the is a C•G base pair at both ends

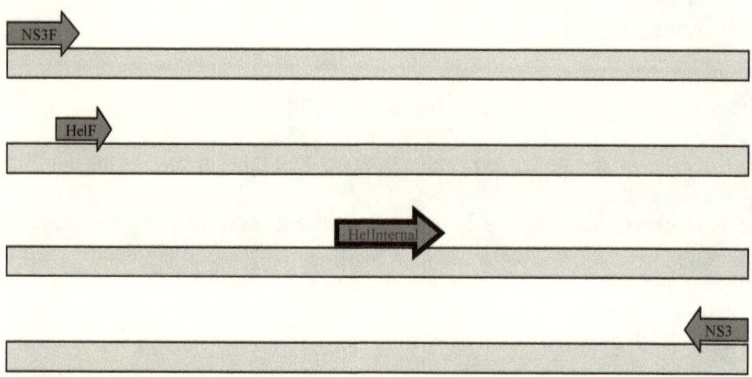

The designed primer sequence was: **gcagagacccatttcctcag**

Figure 68. The NS3 gene and the primers available for it (red). The primer in green is the primer that was manually designed to cover for the sequencing of the center of the helicase's gene.

6.11 Sequencing

All sequencing reactions were sent away to be done. The DNA was purified and concentrated to at least 100ng/μl. At least 10 μl of DNA are required per primer for each sequencing reaction. The primers were supplied with the DNA (prepared primers). Each primer had to be at least 10 μl of 30 μM per sequencing reaction. Sequencing data are summarized in Appendix 6.

6.12 Site Directed Mutagenesis and Primer Design II

Sequencing of the transformed pET vector revealed that both the NS3 domain and the Helicase genes contain a stop codon at the end. There is a TAG triplet in frame that stops the transcription machinery prematurely (see below). The pET vector contains a His-Tag in the 5' position. As a result the His-Tag was out of reading frame. Site Directed Mutagenesis was used to mutate the T base of the TAG triplet to A. This way the specific triplet would not code for a stop codon anymore, but for a Lysine.

The original seq: **cagccggaagaaagTagaagcttgcggcc**
The mutated seq: **cagccggaagaaagAagaagcttgcggcc**

In red are the gene's bases before the stop codon
In blue are the stop codon triplets (top) and the mutated one (bottom)
In green are the vector's bases after the stop codon

Site directed mutagenesis is a PCR based application that involves the extension of a mismatched oligonucleotide, thus incorporating a mutation in a DNA strand that is about to be cloned. The primer designed for SDM was prepared with the criteria that its melting temperature would be well above the 76 °C, that it would span at least 12 bases either way of the base to be replaced and finally that there would be a C≡G base pair rather than an A=T.

6.13 <u>Conclusion</u>

The bacteria culture was optimized to determine the initiation of its lag phase. The optimum culture time under standard conditions is 13 to 14 hours. Stopping the culture before the 13 hours is going to lead in reduced DNA production and leaving the culture for longer than 14 hours will lead to DNA induced with mutations from the death of the bacteria, having entered their lag phase. The transformation protocol and the ligation protocols have both been optimized and are included in the experimental section. Finally it was found that the plasmid pGEM-T is much more efficient and reliable than the initially used pET23-b vector. Although the size of the NS3 domain appeared to be critical for the transformation of cells, after the optimization pure clones-carriers of the NS3 gene were isolated.

A very interesting issue that was also tackled is the issue of choosing either PCR or Bacteria Cloning to amplify the initial genes of the Helicase and the full NS3 domains. Each approach has advantages and disadvantages. The cloning process can be very much compromised especially during ligation. The ligating enzymes have relatively low efficiencies and need large amounts of the gene to work. This is especially true if the desired fragment doesn't have the proper restriction sites and requires some further modifications like blunt-ending. The Helicase and the NS3 genes though behaved fine during their transformation cycles and after a few attempts the protocol was optimised and it worked with high efficacy. The big compromise with PCR is that the reaction is completely reliant on specific DNA primers. Another issue with PCR is the type of polymerase that is used. Some polymerases have poor fidelity, which means that they can induce mutations while copying the target strands. Finally the PCR has a limited base length range that it can cope with and after that the efficacy of the reaction is compromised. The work on the genes though showed that the PCR reaction is better than bacterial cloning when dealing with tiny quantities of DNA. A lot of attention has to be paid though to the selection of the appropriate taq polymerase that is going to be used, since it has to be a proof reading one.

NB. All experimental Data can be found in Appendix 10.

6.14 Gene Sequences & Data for the Dengue Proteins

NS3 DOMAIN

DNA:
AAGCTTgccggagtattgtgggatgttccttcacccccacccatgggaaaggctgaactggaagatggagcctatagaatCaagca
aaaagggattcttggatattcccagatcggagccggagtttacaaagaaggaacattccatacaatgtggcatgtcacacgtggcgct
gtCctaatgcaCaaaggaaagaggattgaaccatcatgggcggacgtcaagaaagaTctaatatcatatggaggaggctggaagt
tagaaggagaatggaaggaaggagaagaagtccaggtattggcaTtggagcctggaaaaaatccaagagccgtccaaacAaaa
cctggtcttttcaGaaccaacgccggaacaataggtgcCgtatctctggacttttctcctggaacgtcaggatctccaatCatTgacaa
aaaaggaaaagttgtgggtctttatggtaatggtgttgttacaaggagtggagcatatgtgagtgctatagcccagacAgaaaaaag
cattgaagacaacccagagatcgaagatgacattttccgaaagagaagactgaccatcatggacctccaTccaggagcgggaaag
acgGaAagataccttccggccatagtcagagaagctataaaacgggtttgagaacattaatcttggcTcccactagagttgtggca
gctgaaatggaggaagcTcttagaggacttccaataagataccaAaccccagccatcagagctgAgcacaccgggcgggagattg
tggacctaatgtgtcatgccacatttaccatgaggctgctatcaccagttagagtgccaaactacaacctgattatcatggacgaagcc
catttcacagacccagcaagtatagcagcCagaggatacatctcaactcgagtggagatgggtgaggcagctgggattttCatgaca
gccactcccccgggaagcagagacccatttcctcagagcaatgcaccaatcatagatgaagaaagagaaatccctgaacgTtcgtg
gaattcTggacatgaGtgggtcacggatttCaaagggaagactgtttggttcgttccaagtataaaagcaggaaatgatatagcagc
ttgcctgagAaaaaatggaaagaaagtgatacaactcagtaggaagacTtttgattctgagtatgtcaagactagaaccaatgattg
ggaTttcgtggttacaactgacatttcagaaatgggtgccaatttcaaggctgagaagggtttatagaccccagacgctgcatgaaacca
gtcatactaacagatggtagGgagcgggtgattctggcaggacctatgccagtgacccactctagtgcagcacaaagaagagggag
aataggaagaaatccaaaaaatgaAaatgaccagtacatatacatgggggaacctctggaaaatgatgaagactgtgcacactgg
aaagaagcCaaaatgctcctagataacatcaacacAccagaaggaatcatCcctagcatgttcgaaccagacgtgaaaaAgtgg
atgccattgatggcgaataccgcttgagaggagaagcaaggaaaacctttgtagacttaatgagaagaggagacctaccagtctggt
tggcctacaAagtggcagctgaaggcatcaactacgcagacagaaggtggtgttttgatggaAtcaGgaacaaccaaatcTtGga
agaaaacgtggaagttgaaatctggacaaaagaaggggaaaggaagaaattgaaacccagatggCtggatgctaggatctattct
gacccactggcgctaaaagaatttaaggaatttgcagccggaagaaagTAGAAGCTT

Protein:
KLAGVLWDVPSPPPMGKAELEDGAYRIKQKGILGYSQIGAGVYKEGTFHTMWHVTRGAVLMHKGKRIEPS
WADVKKDLISYGGGWKLEGEWKEGEEVQVLALEPGKNPRAVQTKPGLFRTNAGTIGAVSLDFSPGTSGSP
IIDKKGKVVGLYGNGVVTRSGAYVSAIAQTEKSIEDNPEIEDDIFRKRRLTIMDLHPGAGKTERYLPAIVREA
IKRGLRTLILAPTRVVAAEMEEALRGLPIRYQTPAIRAEHTGREIVDLMCHATFTMRLLSPVRVPNYNLIIMDE
AHFTDPASIAARGYISTRVEMGEAAGIFMTATPPGSRDPFPQSNAPIIDEEREIPERSWNSGHEWVTDFKG
KTVWFVPSIKAGNDIAACLRKNGKKVIQLSRKTFDSEYVKTRTNDWDFVVTTDISEMGANFKAERVIDPRR
CMKPVILTDGEERVILAGPMPVTHSSAAQRRGRIGRNPKNENDQYIYMGEPLENDEDCAHWKEAKMLLDN
INTPEGIIPSMFEPEREKVDAIDGEYRLRGEARKTFVDLMRRGDLPVWLAYKVAAEGINYADRRWCFDGIR
NNQILEENVEVEIWTKEGERKKLKPRWLDARIYSDPLALKEFKEFAAGRK.KL

Molecular Weight 69916.15 Daltons
622 Amino Acids
93 Strongly Basic(+) Amino Acids (K,R)
89 Strongly Acidic(-) Amino Acids (D,E)
208 Hydrophobic Amino Acids (A,I,L,F,W,V)
114 Polar Amino Acids (N,C,Q,S,T,Y)

8.360 Isolectric Point
5.519 Charge at PH 7.0
Total number of bases translated is 1869
% A = 33.76	[631]	
% G = 26.32	[492]	
% T = 20.39	[381]	
% C = 19.53	[365]	
% Ambiguous = 0.00	[0]	
% A+T = 54.15	[1012]	
% C+G = 45.85	[857]	

BASE COUNT   631 a   365 c   492 g   381 t

HELICASE

DNA:

agactgaccatcatggacctccaTccaggagcgggaaagacgGaAagataccttccggccatagtcagagaagctataaaacggg
gtttgagaacattaatcttggcTcccactagagttgtggcagctgaaatggaggaagcTcttagaggacttccaataagataccaAac
cccagccatcagagctgAgcacaccgggcgggagattgtggacctaatgtgtcatgccacatttaccatgaggctgctatcaccagtt
agagtgccaaactacaacctgattatcatggacgaagcccatttcacagacccagcaagtatagcagcCagaggatacatctcaact
cgagtggagatgggtgaggcagctgggattttCatgacagccactcccccgggaagcagagacccatttcctcagagcaatgcacc
aatcatagatgaagaaagagaaatccctgaacgTtcgtggaattcTggacatgaGtgggtcacggatttCaaagggaagactgttt
ggttcgttccaagtataaaagcaggaaatgatatagcagcttgcctgagAaaaaatggaaagaaagtgatacaactcagtaggaag
acTtttgattctgagtatgtcaagactagaaccaatgattgggaTtttcgtggttacaactgacatttcagaaatgggtgccaatttcaag
gctgagagggttatagacccagacgctgcatgaaaccagtcatctaacagatggtgaGgagcgggtgattctggcaggacctat
gccagtgacccactctagtgcagcacaaagaagagggagaataggaagaaatccaaaaaatgaAaatgaccagtacatatacatg
ggggaacctctggaaaatgatgaagactgtgcacactggaaagaagcCaaaatgctcctagataacatcaacacAccagaaggaa
tcatCcctagcatgttcgaaccagagcgtgaaaaAgtggatgccattgatggcgaataccgcttgagaggagaagcaaggaaaacc
tttgtagacttaatgagaagaggagacctaccagtctggttggcctacaAagtggcagctgaaggcatcaactacgcagacagaag
gtggtgtttgatggaAtcaGgaacaaccaaatcTtGgaagaaaacgtggaagttgaaatctggacaaaagaaggggaaaggaa
gaaattgaaacccagatggCtggatgctaggatctattctgacccactggcgctaaaagaatttaaggaatttgcagccggaagaaa
gTAGAAGCTT

Protein:

RLTIMDLHPGAGKTERYLPAIVREAIKRGLRTLILAPTRVVAAEMEEALRGLPIRYQTPAIRAEHTGREIVDLM
CHATFTMRLLSPVRVPNYNLIIMDEAHFTDPASIAARGYISTRVEMGEAAGIFMTATPPGSRDPFPQSNAPII
DEEREIPERSWNSGHEWVTDFKGKTVWFVPSIKAGNDIAACLRKNGKKVIQLSRKTFDSEYVKTRTNDW
DFVVTTDISEMGANFKAERVIDPRRCMKPVILTDGEERVILAGPMPVTHSSAAQRRGRIGRNPKNENDQYI
YMGEPLENDEDCAHWKEAKMLLDNINTPEGIIPSMFEPEREKVDAIDGEYRLRGEARKTFVDLMRRGDLPV
WLAYKVAAEGINYADRRWCFDGIRNNQILEENVEVEIWTKEGERKKLKPRWLDARIYSDPLALKEFKEFAA
GRK.KL

Molecular Weight 49573.30 Daltons
434 Amino Acids
67 Strongly Basic(+) Amino Acids (K,R)
66 Strongly Acidic(-) Amino Acids (D,E)
147 Hydrophobic Amino Acids (A,I,L,F,W,V)
79 Polar Amino Acids (N,C,Q,S,T,Y)

7.733 Isolectric Point
2.002 Charge at PH 7.0

Total number of bases translated is 1305
% A = 33.95
% G = 26.05
% T = 20.08
% C = 19.92
% Ambiguous 0.00

% A+T = 54.02
% C+T = 40.00

Davis,Botstein,Roth Melting Temp C. 83.37
Wallace Temp C   4330.00

Chapter 7 has been edited by
Dr Chrisanthy P. Vlachaki

CHAPTER 7:

7.1   Introduction

Protein expression is a general term to describe how information encoded in a segment of DNA (a gene) is converted into a functioning protein in a cell. As such, protein express
ion covers the processes of transcription (converting the DNA sequence of a gene into a messenger RNA molecule) and translation (converting the information in the messenger RNA into an amino acid sequence) as well as all of the methods a cell uses to regulate these processes. For example, in a eukaryotic cell, a segment of DNA that comprises a gene is transcribed by an RNA polymerase into a messenger RNA molecule (mRNA) in the nucleus. This mRNA is transported into the cytoplasm of the cell where portions of the mRNA that do not code for amino acids in the protein are removed (in a process called splicing). The "spliced" or mature mRNA is then translated into a string of amino acids by protein "machines" called ribosomes. After translation, the string

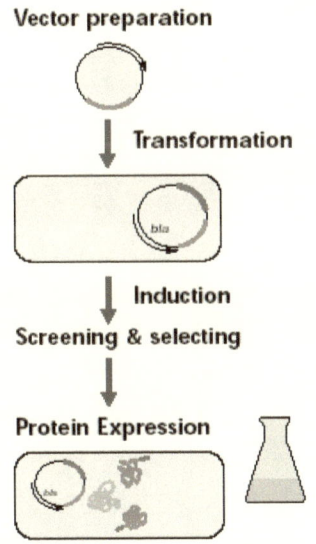

Vector preparation

Transformation

Induction

Screening & selecting

Protein Expression

Figure   69.   Protein Expression

of amino acids folds into a functioning protein that may or may not be modified by adding sugars or other molecules.

Each of these steps from gene to functional protein allow the cells an opportunity to regulate how much of a particular protein is made. Cells can regulate how frequently a particular gene is transcribed into mRNA, how that mRNA is spliced to form a mature mRNA, how long that mRNA lasts in the cell, how frequently the mRNA is translated by ribosomes, how the translated protein is modified and how long the finished protein lasts in the cells.

7.2   the pET expression system

Protein expression systems in E. Coli have been developed and optimized so that even eukaryotic protein will be possible to be expressed. This vector should be capable of producing a bacterial or phage promoter. A very common and representative expression system is the pET Vector system. The pET system is based on the T7 phage promoter (T7 RNA polymerase expression system). T7 polymerase has the unique characteristic that is specific for T7 promoters and will not recognize any DNA from other sources. The latter combined with the facts that those T7 promoters are very rare and that termination sequences for T7 promoters are also rare, make it possible to have very long transcripts with no interruption. Finally it has been found that the T7 RNA polymerase is roughly five times faster than the host (E. Coli) RNA polymerase. As a results genes controlled by the T7 RNA polymerase will be overexpressed.

Special E. Coli strains have been developed that contain the T7 RNA polymerase. The commonest is the BL21 (DE3) strain, which is lysogenic for a specific sequence of the DE3 phage that contains the lacI gene, which contains the lacUV5 promoter, the lacZ gene (beta galactosidase) and the T7 RNA polymerase gene. Responsible for the operation of the T7 RNA polymerase gene is the UV5 promoter. This promoter can be induced by isopropyl beta D thiogalactopyranoside

(IPTG).

So, if the system was perfect the gene-insert should only be expressed in the presence of IPTG, by activation of the lacUV5 promoter that will eventually switch on the T7 RNA polymerase. In practice the E. Coli polymerase can produce tiny amounts of T7 RNA polymerase without IPTG. These small amounts of T7 RNA polymerases may then bind to the lacUV5 promoter and initiate gene expression, without IPTG induction. In an attempt to eliminate the "background" expression of the gene before induction the T7 lysozyme has been introduced as a natural inhibitor of the T7 RNA polymerase. The lysozyme is used primarily as a defense agent by the cell. The lysozyme can attack the cell wall, but only from its outer surface. As a result it cannot harm its host but can destroy any invaders. The gene of the T7 lysozyme is expressed on a second plasmid and as a result the concentrations of T7 RNA polymerase are suppressed. There are two different plasmids for expressing the T7 lysozyme. The first is called pLysE and is expressed from the tet promoter, whereas the second one, known as pLysS is oriented the other way. The difference between the two is that the pLysE can produce more T7 lysozyme, but affects the growth of the cells. On the other hand the pLysS does not affect the growth rate of the cells but does not produce as high quantities of T7 lysozyme. Of course, after IPTG induction the T7 RNA polymerase concentration will be bigger than the T7 lysozyme. The whole gain from this system is that even very toxic proteins can be produced. Because the background expression can kill the cells before induction, a further optimization was made for the case of extremely toxic proteins. The lac operator was positioned upstream the lacUV5 promoter and the lacI was incorporated to the pET system (the lacI will produce a lac repressor that will then bind to the operator). As a result the T7 RNA polymerase will not be able to read the T7 polymerase gene. IPTG induction will solve this because the lacUV5 promoter is further down the operator-repressor complex.

7.3   Protein Expression

The bacteria of the overnight cultures will have definitely reached the stationary phase of their growth cycle. As a result those bacteria will not be interested in growing, they should have low metabolism and reduced numbers of ribosomes. After the inoculation in the fresh medium it is expected to take more to the bacteria to adjust. So, the lag phase is expected to be longer than usual.

The lag phase is the adjustment phase. During this period, the cells adjust to their new environment either passively (adjust to pH, nutrients and temperature) or actively (acquire antibiotic resistance). There is no significant growth of the culture during the lag phase. Following the lag phase is the exponential phase. The adjustment of the bacteria has been completed and bacteria increase fast in numbers. With the nutrient concentration declining and the secondary metabolite & waste concentrations increasing, the rate of increase in bacterial numbers will soon drop and eventually the bacterial numbers will remain constant over time. The growth rate will equal the death rate at this point. Finally the death phase will follow. Bacteria numbers will decline in this phase, since the death rate is bigger than the growth rate. Usually, the death phase is due to unavailability of nutrients and increased toxicity of the suspension.

Freshly growing bacteria will start to grow in the right conditions (correct medium and temperature). The time it will take though for those bacteria to reach their max log rate of growth depends on a variety of parameters with the most important ones being the species of the bacteria, the type and composition of the medium and the oxygen availability. The strain used for the expression of both proteins was the BL21 E.Coli species. Those cells were transformed with the vector pET23c containing the desired gene in reading frame. The BL21lys cells were transformed under standard transformation procedures. The plasmids were inserted into the cells using the heat-shock transformation approach. For every protein expression experiment a new batch of transformed bacteria plates must be prepared. 500 ml of LB was prepared, containing the ampicilin (1 μl/ml) and chloramphenicol (1 μl/ml) antibiotics. The combination of the two antibiotics eliminates the possibilities of contamination of the cultures with foreign bacterial species. The 500 ml LB flasks were kept at 37°C overnight in an incubator. Another 50 ml of LB with ampicillin and chloramphenicol (1 μl/ml) were prepared and they were inoculated with single colonies picked up from freshly prepared agar (LB+AMP+Chlor) plates. The colony-inoculated 50ml suspensions were kept in a shacking incubator at 37 °C and 300 rpm. The suspensions were incubated for 18 hours. 5 ml of each suspension was inoculated in each of the 500 ml suspensions.

A spectrophotometer was calibrated against plain LB at 600 nm. Measurements of the optical density of each culture were taken in half an hour intervals, until the optical density of each culture reached 0.4. When it the OD becomes 0.4 the cell numbers are optimal for induction. The induction was used using I-PDG (1 M) inducer for the pET23c system. One, two and three hours after the induction 1 ml of each cell suspension were taken and prepared for SDS-Page.

SDS-Page (sodium dodecyl (lauryl) sulfate-polyacrylamide gel electrophoresis) is a method designed for running and separating proteins. With an SDS-Page it is possible to establish the protein size, to identity the protein, to determine its purity, to identify disulfide bonds and to quantify the amount of protein in its sample. Finally, SDS-Page is used for various blotting purposes.

Sodium dodecyl sulfate is a detergent that will denature the protein from its quaternary structure down to its primary (linear). The reason for the need to denature the protein is to be able to compare and correlate the resulting bands to its other after running the gel. Since the SDS-Page separates the protein accordingly to their size, and proteins have to go through a mesh of acrylamide, proteins of the same molecular weight with different shape would result in two different bands. SDS is an anionic detergent that will bind quantitatively to proteins thus denaturing them. The high negative charge of the SDS is much higher than the charge any protein may have. The hydrophobic tail of the SDS can establish strong interaction with protein chains (polypeptide). The quantity of SDS molecules that will interact with a given protein depends on the number of amino acids of this protein. The charge of the protein is not playing a very significant role if the fact that each SDS molecule will contribute two negative charges. An electric current is the force that will move the proteins on a polyacrylamide gel. The negative protein molecules will move towards the positive end. For the fixation of the proteins 25% acetic acid in water was used, since it will keep the proteins denatured. The SDS-Page gel was stained with Coomasie blue dye R250. Both the fixative and the dye were prepared in the same solution with methanol as a solvent. Finally, the gel is was destained from the Coomasie blue dye and dried.

Preparing cell suspension samples for SDS Page involves the spinning down of the 1 ml suspension in an eppendorf and its re-suspension in 1x SDS buffer. The buffer was prepared by adding 500 µl of 5x SDS loading dye, 50 µl of B-Mercapetanol and 1950 µl of Sigma H_2O. The quantity of 1x SDS loading buffer, in which each cell suspension sample was to be resuspended, depended on the OD of that sample. For each OD unit, 100 μl of 1x SDS loading buffer were added.

After taking 3 samples from each suspension (1,2 and 3 hours), an SDS-Page gel was prepared according to the protocol in Appendix 10. 8 μl of cells in SDS dye were loaded in each lane along 12 μl of marker on either sides of the gel. The results of the SDS-Page showed that there is protein expression (figures 71-72). That was only for the helicase though. The NS3 protein was not showing at all. This can only mean two things: either that there is no protein production or the protein is being produced, but it is shadowed by other cellular proteins of the same size. The only way to confirm protein production is to run a western blot. That was not possible though since a problem occurred in the last triplet of the gene. A stop codon was inserted and the His tag was out of reading frame. This was fixed by site-directed mutagenesis.

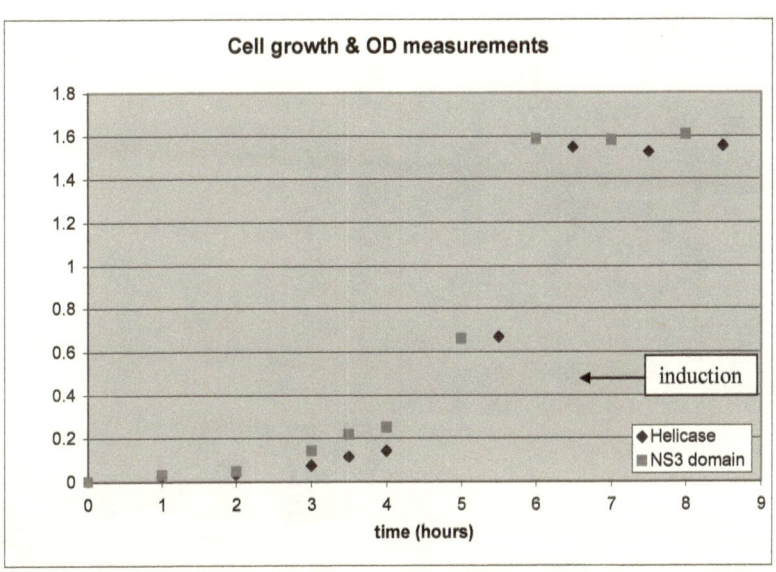

Figure 70. Top: Cell growth of the protein expression cultures – Optical density (OD) versus time. The lag phases of both cultures last approximately 2 to 3 hours, which is too long for lag phase. Then between the 3rd and the 6th hour there is the exponential phase and finally it is clear that approximately after 6th hour both cultures enter their stationary phases.

Right: Typical bacterial suspension growth cycle. The culture starts with the lag phase, also known as the adjustment phase, moves on to the exponential phase, which is the actual cell growth phase. At this stage the growth rate is higher than the death rate. The next stage is the stationary stage, where the growth rate is equal to the death rate. Finally the last phase is the death phase, where the death rate id higher than the growth rate.

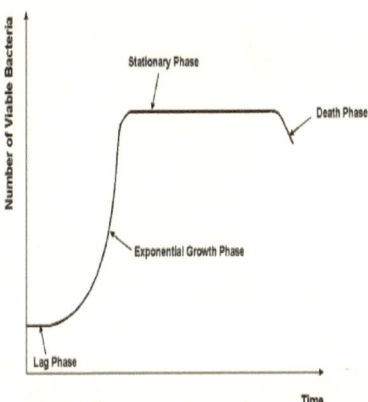

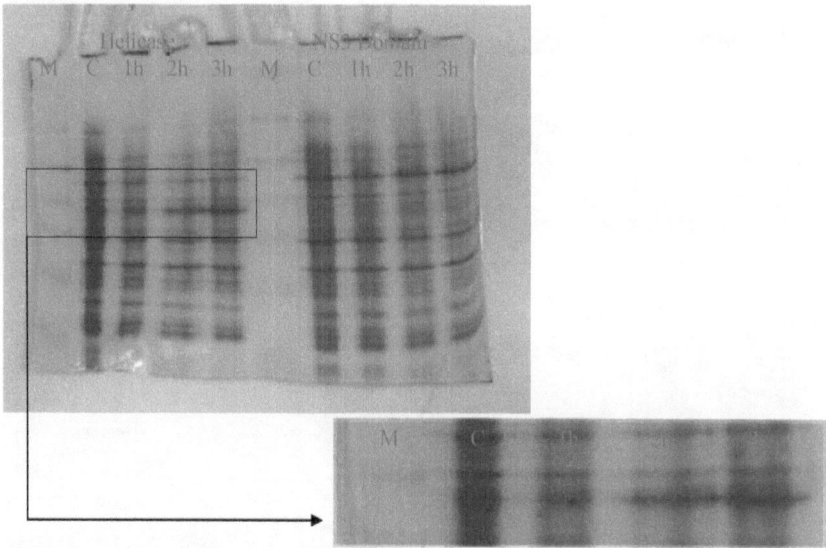

Figure 71. The SDS-gel from the protein expression experiment. On top it is clear that there is protein expression, since the band gets stronger with time. The helicase is estimated to be approx. 52 KDalton whereas the NS3 domain should be around 78 KDalton

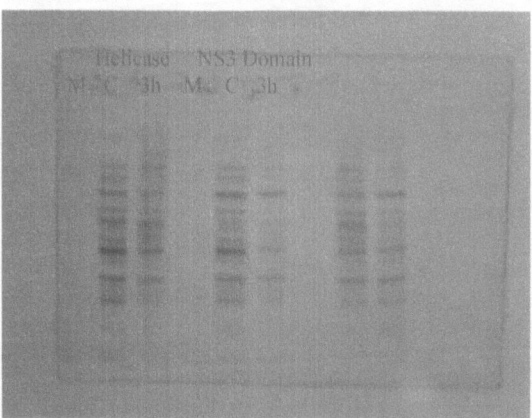

Figure 72. The expression was repeated again and after 3 hours the bands did not show any stronger than the control. The proteins may be hidden under cellular proteins of the same size.

7.4   Conclusions

The levels of protein productions are not clear, unless a western blot is done on the SDS-PaGE gels. Unfortunatelly, running a western blot requires the His-Tag of the pET vector to be in reading frame, and in this particular experiment (as has already been mentioned in the previous chapter) is not. All due to a stop codon found in the very last triplet of the coding gene. Although the band of the NS3 domain is not clear, this does not mean that the protein is not there. The chances are that the NS3 protein complex is produced at higher levels than the Helicase protein. This is based on the OD doubling time of the two cultures. It appears that the Helicase is a bit tougher on the cells and it takes a bit longer to the cells to adjust to its production. After site-directed mutagenesis of one base of the stop triplet everything should shift into frame and running the western blot with specific antibodies will reveal the actual amounts of the Helicase and the NS3 protein complexes being produced.

References

1. Purves, Orians, Heller, Sadava, (1998) Life, the science of Biology, 3rd ed., Sinauer, Freeman.

2. Allan Jones, Rob Reed and Jonathan Weyers (1998) Practical Skills in Biology (second edition). Published by Longman. ISBN 0-582-29885-7

3. Prescott (1998) Microbiology, 3rd edition, WYB Publishers, USA

4. Rob Reed et al. (1998) Practical Skills in Biomolecular Sciences (first edition). Published by Longman. ISBN 0-582-29826-1.

5. Schaum's (1999) Schaum's outlines Biochemistry (second edition). Published by Mc Graw Hill Publications. ISBN 0-07-03614905.

6. Mathews et al. (2000) Biochemistry (third edition). Published by Addison Wesley, Longman. ISBN 0-8053-3066-6.

7. Geoffrey L. Zubay (1998) Biochemistry (forth edition). Published by Mc Graw Hill Publications. ISBN 0-697-21900-3.

8. L M Prescot, J P Harley and D A Klein, *Microbiology*, 4th ed. 1999.

9. Glen L. Prosise, (2002),Crystal Structure of Tritrichomonas foetus Inosine Monophosphate.*THE JOURNAL OF BIOLOGICAL CHEMISTRY*.**277**,50654-50659.

10.   Carbó-Dorca R and Besalú E (2000),Quantum theory of QSAR. *CONTRIBUTIONS to SCIENCE*.**1**,399-422.

11.   Neddermann, P., Tomei, L., Steinkuhler, C., Gallinari, P., Tramontano,A. & De Francesco, R. (1997). The nonstructural proteins of thehepatitis C virus: structure and functions. *Biol. Chem.* **378**, 469–476.

12.   Failla, C., Tomei, L. & De Francesco, R. (1995). An amino-terminaldomain of the hepatitis C virus NS3 protease is essential forinteraction with NS4A. *J. Virol.* **69**, 1769–1777.

13.   Bartenschlager, R., Lohmann, V., Wilkinson, T. & Koch, J.O. (1995).Complex formation between the NS3 serine-type proteinase of thehepatitis C virus and NS4A and its importance for polyproteinmaturation. *J. Virol.* **69**, 7519–7528.

14.   Tanji, Y., Hijikata, M., Satoh, S., Kaneko, T. & Shimotohno, K. (1995).Hepatitis C virus-encoded nonstructural protein NS4A has versatilefunctions in viral protein processing. *J. Virol.* **69**, 1575–1581.

15.   Kim, J.L., *et al.*, & Thomson, J.A. (1996). Crystal structure of thehepatitis C virus NS3 protease domain complexed with a syntheticNS4A cofactor peptide. *Cell* **87**, 343–355.

16.   Markland, W., *et al.*, & Chambers, S.P. (1997). Purification andcharacterization of the NS3 serine protease domain of hepatitis C virus expressed in *Saccharomyces cerevisiae. J. Gen. Virol.* **78**, 39–43.

17.   Steinkuhler, C., *et al.*, & De Francesco, R. (1996). Activity of purified hepatitis C virus protease NS3 on peptide substrates. *J. Virol.* **70**,6694–6700.

18.   Suzich, J.A., *et al.*, & Collett, M.S. (1993). Hepatitis C virus NS3protein polynucleotide-stimulated nucleoside triphosphatase and comparison with the related pestivirus and flavivirus enzymes. *J. Virol.***67**, 6152–6158.

19.   Tai, C.L., Chi, W.K., Chen, D.S. & Hwang, L.H. (1996). The helicase activity associated with hepatitis C virus nonstructural protein 3(NS3). *J. Virol.* **70**, 8477–8484.

20.   Jin, L. & Peterson, D.L. (1995). Expression, isolation, and characterization of the hepatitis C virus ATPase/RNA helicase. *Arch.Biochem. Biophys.* **323**, 47–53.

21.   Preugschat, F., Averett, D.R., Clarke, B.E. & Porter, D.J.T. (1996). A steady-state and pre-steady-state kinetic analysis of the NTPase activity associated with the hepatitis C virus NS3 helicase domain. *J. Biol. Chem.* **271**, 24449–24457.

22.   Morgenstern, K.A., *et al.*, & Thomson, J.A. (1997). Polynucleotide modulation of the protease, nucleoside triphosphatase, and helicase activities of a hepatitis C virus NS3–NS4A complex isolated from transfected COS cells. *J. Virol.* **71**, 3767–3775.

23.   Hong, Z., *et al.*, & Kwong, A.D. (1996). Enzymatic characterization of hepatitis C virus NS3/4A complexes expressed in mammalian cells by using the herpes simplex virus amplicon system. *J. Virol.* **70**,4261–4268.

24.   Gwack, Y., Kim, D.W., Han, J.H. & Choe, J. (1996). Characterization of RNA binding activity and RNA helicase activity of the hepatitis C virus NS3 protein. *Biochem. Biophys. Res. Commun.* **225**, 654–659.

25.   Kanai, A., Tanabe, K. & Kohara, M. (1995). Poly(U) binding activity of hepatitis C virus NS3 protein, a putative RNA helicase. *FEBS Lett.* **376**, 221–224.

26.   Miller, R.H. & Purcell, R.H. (1990). Hepatitis C virus shares amino acid sequence similarity with pestiviruses and flaviviruses as well as members of two plant virus supergroups. *Proc. Natl. Acad. Sci. USA* **87**, 2057–2061.

27.   Kadare, G. & Haenni, A.L. (1997). Virus-encoded RNA helicases. *J. Virol.* **71**, 2583–2590.

28.   Mirzayan, C. & Wimmer, E. (1992). Genetic analysis of an NTPbinding motif in poliovirus polypeptide 2C. *Virology* **189**, 547–555.

29.   MacPherson, P., Thorner, L., Parker, L.M. & Botchan, M. (1994). The bovine papilloma virus E1 protein has ATPase activity essential to viral DNA replication and efficient transformation in cells. *Virology* **204**, 403–408.

30.   Martinez, R., Shao, L. & Weller, S.K. (1992). The conserved helicase motifs of the herpes simplex virus type 1 origin-binding protein UL9 are important for function. *J. Virol.* **66**, 6735–6746.

31.   Kolykhalov, A.A., Agapov, E.V., Blight, K.J., Mihalik, K., Feinstone, S.M. & Rice, C.M. (1997). Transmission of hepatitis C by intrahepatic inoculation with transcribed RNA. *Science* **277**, 570–574.

32.   Gorbalenya, A. & Koonin, E.V. (1993). Helicases: amino acid sequence comparisons and structure-function relationships. *Curr. Opin. Struct. Biol.* **3**, 419–429.

33.   Lohman, T.M. & Bjornson, K.P. (1996). Mechanisms of helicasecatalyzed DNA unwinding. *Annu. Rev. Biochem.* **65**, 169–214.

34.   Yao, N., Hesson, T., Cable, M., Hong, Z., Kwong, A.D., Le, H.V. & Weber, P.C. (1997). Structure of the hepatitis C virus RNA helicase domain. *Nat. Struct. Biol.* **4**, 463–467.

35.   Subramanya, H.S., Bird, L.E., Brannigan, J.A. & Wigley, D.B. (1996). Crystal structure of a DExx box DNA helicase. *Nature* **384**, 379–383.

36.   Korolev, S., Hsieh, J., Gauss, G.H., Lohman, T.M. & Waksman, G. (1997). Major domain swiveling revealed by the crystal structures of complexes of *E. coli* Rep helicase bound to single-stranded DNA and ADP. *Cell* **90**, 635–647.

37.   Bartenschlager, R., Ahlborn-Laake, L., Mous, J. & Jacobsen, H. (1994). Kinetic and structural analyses of hepatitis C virus polyprotein processing. *J. Virol.* **68**, 5045–5055.

38.   Walker, J.E., Saraste, M., Runswick, M.J. & Gay, N.J. (1982). Distantly related sequences in the alpha- and beta-subunits of ATP synthase, myosin, kinases and other ATP-requiring enzymes and a common nucleotide binding fold. *EMBO J.* **1**, 945–951.

39.   Saraste, M., Sibbald, P.R. & Wittinghofer, A. (1990). The P-loop – a common motif in ATP- and GTP-binding proteins. *Trends Biochem. Sci.* **15**, 430–434.

40.   Raghunathan, S., Ricard, C.S., Lohman, T.M. & Waksman, G. (1997). Crystal structure of the homo-tetrameric DNA-binding domain of *Escherichia coli* single-stranded DNA-binding protein determined by multiwavelength X-ray diffraction on the selenomethionyl protein at 2.9 Å resolution. *Proc. Natl. Acad. Sci. USA* **94**, 6652–6657.

41.   Ruff, M., *et al.*, & Moras, D. (1991). Class II aminoacyl transfer RNA synthetases: crystal structure of yeast aspartyl-tRNA synthetase complexed with tRNA(Asp). *Science* **252**, 1682–1689.

42.   Bochkarev, A., Pfuetzner, R.A., Edwards, A.M. & Frappier, L. (1997). Structure of the single-stranded-DNA-binding domain of replication protein A bound to DNA. *Nature* **385**, 176–181.

43.   Kim, Y., Geiger, J.H., Hahn, S. & Sigler, P.B. (1993). Crystal structure of a yeast TBP/TATA-box complex. *Nature* **365**, 512–520.

44.   Kim, J.L., Nikolov, D.B. & Burley, S.K. (1993). Co-crystal structure of TBP recognizing the minor groove of a TATA element. *Nature* **365**, 520–527.

45.   George, J.W., Brosh, R.M., Jr. & Matson, S.W. (1994). A dominant negative allele of the *Escherichia coli* uvrD gene encoding DNA helicase II. A biochemical and genetic characterization. *J. Mol. Biol.* **235**, 424–435.

46.   Seeley, T.W. & Grossman, L. (1990). The role of *Escherichia coli* UvrB in nucleotide excision repair. *J. Biol. Chem.* **265**, 7158–7165.

47.   Black, M.E. & Hruby, D.E. (1992). Site-directed mutagenesis of a conserved domain in vaccinia virus thymidine kinase. Evidence for a potential role in magnesium binding. *J. Biol. Chem.* **267**, 6801–6806.

48.   Yan, H.G. & Tsai, M.D. (1991). Mechanism of adenylate kinase. Demonstration of a functional relationship between aspartate 93 and Mg2+ by site-directed mutagenesis and proton, phosphorus-31, and magnesium-25 NMR. *Biochemistry* **30**, 5539–5546.

49.   Gross, C.H. & Shuman, S. (1995). Mutational analysis of vaccinia virus nucleoside triphosphate phosphohydrolase II, a DExH box RNA helicase. *J. Virol.* **69**, 4727–4736.

50.   Brosh, R.M., Jr. & Matson, S.W. (1995). Mutations in motif II of *Escherichia coli* DNA helicase II render the enzyme nonfunctional in both mismatch repair and excision repair with differential effects on the unwinding reaction. *J. Bacteriol.* **177**, 5612–5621.

51.   Pause, A. & Sonenberg, N. (1992). Mutational analysis of a DEAD box RNA helicase: the mammalian translation initiation factor eIF-4A.*EMBO J.* **11**, 2643–2654.

52.   Heilek, G.M. & Peterson, M.G. (1997). A point mutation abolishes the helicase but not the nucleoside triphosphatase activity of hepatitis Cvirus NS3 protein. *J. Virol.* **71**, 6264–6266

53.   Gross, C.H. & Shuman, S. (1996). The QRxGRxGRxxxG motif of the vaccinia virus DExH box RNA helicase NPH- II is required for ATP hydrolysis and RNA unwinding but not for RNA binding. *J. Virol.* **70**, 1706–1713.

54.   Pause, A., Methot, N. & Sonenberg, N. (1993). The HRIGRXXR region of the DEAD box RNA helicase eukaryotic translation initiation factor 4A is required for RNA binding and ATP hydrolysis. *Mol. Cell Biol.* **13**, 6789–6798.

55.   Hall, M.C. & Matson, S.W. (1997). Mutation of a highly conserved arginine in motif IV of *Escherichia coli* DNA helicase II results in an ATP-binding defect. *J. Biol. Chem.* **272**, 18614–18620.

56.   Bilderback, T., Fulmer, T., Mantulin, W.W. & Glaser, M. (1996). Substrate binding causes movement in the ATP binding domain of *Escherichia coli* adenylate kinase. *Biochemistry* **35**, 6100–6106.

57.   Schulz, G.E. (1992). Induced-fit movements in adenylate kinases. *Faraday Discuss.* **93**, 85–93.

58.   Moore, K.J. & Lohman, T.M. (1994). Kinetic mechanism of adenine nucleotide binding to and hydrolysis by the *Escherichia coli* Rep monomer. 1. Use of fluorescent nucleotide analogues. *Biochemistry* **33**, 14550–14564.

59.   Chao, K. & Lohman, T.M. (1990). DNA and nucleotide-induced conformational changes in the *Escherichia coli* Rep and helicase II (UvrD) proteins. *J. Biol. Chem.* **265**, 1067–1076.

60.   Hakansson, K., Doherty, A.J., Shuman, S. & Wigley, D.B. (1997). X-ray crystallography reveals a large conformational change during guanyl transfer by mRNA capping enzymes. *Cell* **89**, 545–553.

61.   Subramanya, H.S., Doherty, A.J., Ashford, S.R. & Wigley, D.B. (1996). Crystal structure of an ATP-dependent DNA ligase from bacteriophage T7. *Cell* **85**, 607–615.

62.   Phillips, R.J., Hickleton, D.C., Boehmer, P.E. & Emmerson, P.T. (1997). The RecB protein of *Escherichia coli* translocates along singlestranded DNA in the 3 to 5 direction: a proposed ratchet mechanism. *Mol. Gen. Genet.* **254**, 319–329.

63.   Ali, J.A. & Lohman, T.M. (1997). Kinetic measurement of the step size of DNA unwinding by *Escherichia coli* UvrD helicase. *Science* **275**, 377–380.

64.   Chen, Y.Z., Zhuang, W. & Prohofsky, E.W. (1992). Energy flow considerations and thermal fluctuational opening of DNA base pairs at a replicating fork: unwinding consistent with observed replication rates. *J. Biomol. Struct. Dyn.* **10**, 415–427.

65.   Love R A et al. (1997), Conformational Changes in Hepatitis C Virus NS3 Proteinase due toNS4A Cofactor Complexation. *Experimental Progress Reports*. **7**,102-104.

66.   Casadio R, et al. (2001),Computer Aided Design of peptidomimetic molecules active on vascular cell growth and migration. *Chemistry*. **1**,102-109.

67.   TAI C L et al. (1996),The Helicase Activity Associated with Hepatitis C Virus Nonstructural Protein 3 (NS3). *JOURNAL OF VIROLOGY.* **70(12)**,8477-8484.

68.   Josheph L Kim et al (1998),Hepatitis C virus NS3 RNA helicase domain with a bound oligonucleotide: the crystal structure provides insights into the mode of unwinding. *Structure.***6**,89-100.

69.   KIM D W et al. (1997),Mutational Analysis of the Hepatitis C Virus RNA Helicase. *JOURNAL OF VIROLOGY.* **71(12)**,9400-9409.

70.   Locatelli et al (2002),Hepatitis C virus NS£ ATPase/Helicase: An ATP Switch Regulates the Cooperativity among different Substrate Binding Sites. *Biochemistry.* **41**,10332-10342.

71.   Michael MC Lai (2001),RNA polymerase as an antiviral target of hepatitis C virus. *Antiviral Chemistry & Chemotherapy.* **12**,143-147.

72.   Dymock et al (2000),Novel approaches to the treatment of hepatitis C virus infection. *Antiviral Chemistry & Chemotherapy.* **11**,79-96.

73.   Cho H S et al. (1998),Crystal Structure of RNA Helicase from Genotype 1b Hepatitis C Virus. *THE JOURNAL OF BIOLOGICAL CHEMISTRY.* **273(24)**,15045-15052.

74.   Clarke B (1997), Molecular virology of hepatitis C virus. *Journal of General Virology.* **78**,2397-2410.

75.   Caruthers et al (2002),Helicase structure and mechanism. *Current Opinion in Structural Biology.* **12**,123-133.

76.   Martin F et al. (1998), Design of Selective Eglin Inhibitors of HCV NS3 Proteinase. *Biochemistry.* **37**,11459-11468.

77.   Watson J D and Crick F H (1953), Molecular Structure of Nucleic Acids. *Nature.* **171**,737-738.

78.   Hsu C C et al. (1998), An ELISA for RNA Helicase Activity: Application as an Assay of the NS3 Helicase of Hepatitis C Virus. *BIOCHEMICAL   AND   BIOPHYSICAL   RESEARCH COMMUNICATIONS.* **253**,594-599.

79.   Morris P D et al. (2002),Hepatitis C Virus NS3 and Simian Virus 40 T Antigen Helicases Displace Streptavidin from 5¢-Biotinylated Oligonucleotides but Not from 3¢-Biotinylated Oligonucleotides: Evidence for Directional Bias in Translocation on Single-Stranded DNA. *Biochemistry.***41**,2372-2378.

80.   Tessmann K et al. (2002),Cloning and molecular characterization of human highaffinity antibody fragments against Hepatitis C virus NS3 helicase. *Journal of Virological Methods.***103**,75-88.

81.   Pento´n N et al. (2002),Antigenicity of a recombinant NS3 protein representative of ATPase/Helicase domain from Hepatitis C virus. *Clinical Biochemistry.***35**,0.

82.   Borowski P et al. (2002), Nucleotide triphosphatase /helicase of hepatitis C virus as a target for antiviral therapy. *Antiviral Research.***55**,397-412.

83.   RHO J et al. (2001),The Arginine-1493 Residue in QRRGRTGR1493G Motif IV of the Hepatitis C Virus NS3 Helicase Domain Is Essential for NS3 Protein Methylation by the Protein Arginine Methyltransferase 1. *JOURNAL OF VIROLOGY.* **75(17)**,8031-8044.

84.   Liu D et al. (2001),Solution Structure and Backbone Dynamics of an Engineered Arginine-rich Subdomain 2 of the Hepatitis C Virus NS3 RNA Helicase. *J. Mol. Biol.* **314**,543-561.

85.   Giada Aurora Locatelli, Silvio Spadari, and Giovanni Maga (2002),Hepatitis C Virus NS3 ATPase/Helicase: An ATP Switch Regulates the Cooperativity among the Different Substrate Binding Sites. *Biochemistry*.**41**,10332-10342.

86.   Brown L M et al. (2002),A single amino acid is critical for the expression of B-cell epitopes on the helicase domain of the pestivirus NS3 protein. *Virus Research*. **84**,111-124.

87.   Aoubala M et al. (2001),The inhibition of cAMP-dependent protein kinase by full-length hepatitis C virus NS3/4A complex is due to ATP hydrolysis. *Journal of General Virology*. **82**,1637-1646.

88.   Pang P S et al. (2002),The hepaptitis C viral NS3 protein is a processive   DNA helicase with cofactor enhanced RNA unwinding. *The EMBO Journal*. **21(5)**,1168-1176.

89.   Johansson et al (2001),Inhibition of Hepatitis C Virus NS3 Protease Activity by Product-Based Peptides is Dependent on Helicase Domain. *Bioorganic & Medicinal Chemistry Letters*. **11**,203-206.

90.   Phoon C W et al. (2001), Biological Evaluation of Hep C Virus Helicase Inhibitors. *Bioorganic & Medicinal Chemistry Letters*. **11**,1647-1650.

91.   TAI C L et al. (2001),Structure-Based Mutational Analysis of the Hepatitis C Virus NS3 Helicase. *Journal of Virology*. **75**,8289-8297.

92.   Alaoui-Ismaili M H et al. (2000),A novel high throughput screening assay for HCV NS3 helicase activity. *Antiviral Research*. **46**,181-193.

93.   Sarver RW et al (2002),Physical methods to determine the binding mode of putative ligands for hepatitis C virus NS3 helicase. *Analytical Biochemistry*. **309**,186-195.

94.   Ligand-Protein Contacts (LPC) were derived with LPC software (Sobolev V., Sorokine A., Prilusky J., Abola E.E. and Edelman M. (1999) Automated analysis of interatomic contacts in proteins. *Bioinformatics*, **15**, 327-332).

95.   Contacts of Structural Units (CSU) were derived with CSU software (Sobolev V., Sorokine A., Prilusky J., Abola E.E. and Edelman M. (1999) Automated analysis of interatomic contacts in proteins. *Bioinformatics*, **15**, 327-332).

96.   SwissProt SPDBV DeepViewer, ver. 3.7, Nicolas Guex *et al*. GlaxoSmithKline, http://www.expasy.org/spdbv

97.   ClustalW,   (Higgins   and   Sharp,   1989), http://www.ebi.ac.uk/clustalw

98.   Threading (1997), *BioInformatics*. **13**, 345-356

99.   HyperChem © (2000) HypeCube, version 7.

100.   Molecular Operating Environment © (2002) Chemical Computing Group, version 2002.03.

101.   PyMol © (2001) Delano Scientific, version 2.1

102.   LigBuilder (2001) Institute of Physical Chemistry, Peking University, China, version 1.2

103.   ChemOffice Ultra ® (2001) CambridgeSoft, version 7.01

104.   ChemSketch Freeware (2002) Advanced Chemistry Development, version 5.12

105.   Sybyl Molecular suite (v6.9) Tripos Software, Cambridge

106.   Alter, M. J., Kruszon-Moran, D., Nainan, O. V., McQuillan, G. M., Gao, F., Moyer, L. A., Kaslow, R. A., and Margolis, H. S. (1999) *N. Engl. J. Med.* **341,** 556-562

107.   Frick, D. N. (2003) *Curr. Org. Chem.* **7,** in press

108.   Leveque, V. J., and Wang, Q. M. (2002) *Cell. Mol. Life Sci.* **59,** 909–919

109.   Narjes, F., Koch, U., and Steinkuhler, C. (2003) *Exp. Opin. Investig. Drugs* **12,** 153–163

110.   Pang, P. S., Jankowsky, E., Planet, P. J., and Pyle, A. M. (2002) *EMBO J.* **21,** 1168–1176

111.   Yao, N., Hesson, T., Cable, M., Hong, Z., Kwong, A. D., Le, H. V., and Weber, P. C. (1997) *Nat. Struct. Biol.* **4,** 463–467

112.   Cho, H. S., Ha, N. C., Kang, L. W., Chung, K. M., Back, S. H., Jang, S. K., and Oh, B. H. (1998) *J. Biol. Chem.* **273,** 15045–15052

113.   Kim, J. L., Morgenstern, K. A., Griffith, J. P., Dwyer, M. D., Thomson, J. A., Murcko, M. A., Lin, C., and Caron, P. R. (1998) *Structure* **6,** 89–100

114.   Yao, N., Reichert, P., Taremi, S. S., Prosise, W. W., and Weber, P. C. (1999) *Structure Fold Des.* **7,** 1353–1363

115.   Liu, D., Wang, Y. S., Gesell, J. J., and Wyss, D. F. (2001) *J. Mol. Biol.* **314,** 543–561

116.   Gesell, J. J., Liu, D., Madison, V. S., Hesson, T., Wang, Y. S., Weber, P. C., and Wyss, D. F. (2001) *Protein Eng.* **14,** 573–582

117.   Gorbalenya, A. E., and Koonin, E. V. (1993)   *Curr. Opin. Struct. Biol.* **3,** 419–429

118.   Kwong, A. D., Kim, J. L., and Lin, C. (2000) *Curr. Top Microbiol. Immunol.* **242,** 171–196

119.   Lam, A. M. I., Keeney, D., Eckert, P. Q., and Frick, D. N. (2003) *J. Virol.* **77,** 3950–3961

120.   Yanagi, M., Purcell, R. H., Emerson, S. U., and Bukh, J. (1997) *Proc. Natl. Acad. Sci. U. S. A.* **94,** 8738-8743

121.   Dubendorff, J. W., and Studier, F. W. (1991) *J. Mol. Biol.* **219,** 45-59

122.   Morris, P. D., and Raney, K. D. (1999) *Biochemistry* **38,** 5164-5171

123.   Morris, P. D., Byrd, A. K., Tackett, A. J., Cameron, C. E., Tanega, P., Ott, R., Fanning, E., and Raney, K. D. (2002) *Biochemistry* **41,** 2372-2378

124.   Thompson, J. D., Gibson, T. J., Plewniak, F., Jeanmougin, F., and Higgins, D. G. (1997) *Nucleic Acids Res.* **25,** 4876-4882

125.   Chamberlain, R. W., Adams, N., Saeed, A. A., Simmonds, P., and Elliott, R. M. (1997) *J. Gen. Virol.* **78,** 1341-1347

126.   Lin, C., and Kim, J. L. (1999) *J. Virol.* **73,** 8798-8807

127.   Preugschat, F., Danger, D. P., Carter, L. H., 3rd, Davis, R. G., and Porter, D. J. (2000) *Biochemistry* **39,** 5174-5183

128.   Paolini, C., Lahm, A., De Francesco, R., and Gallinari, P. (2000) *J. Gen. Virol.* **81,** 1649-1658

129.   Tai, C. L., Pan, W. C., Liaw, S. H., Yang, U. C., Hwang, L. H., and Chen, D. S. (2001) *J. Virol.* **75,** 8289-8297

130.   Kim, J. W., Seo, M. Y., Shelat, A., Kim, C. S., Kwon, T. W., Lu, H. H., Moustakas, D. T., Sun, J., and Han, J. H. (2003) *J. Virol.* **77,** 571-582

131.   Caruthers, J. M., Johnson, E. R., and McKay, D. B. (2000) *Proc. Natl. Acad. Sci. U. S. A.* **97,** 13080-13085

132.   Korolev, S., Yao, N., Lohman, T. M., Weber, P. C., and Waksman, G. (1998) *Protein Sci* **7,** 605-610

133.   Soultanas, P., Dillingham, M. S., Wiley, P., Webb, M. R., and Wigley, D. B. (2000) *EMBO J.* **19,** 3799-3810

134.   Porter, D. J., and Preugschat, F. (2000) *Biochemistry* **39,** 5166–5173

135.   Gwack, Y., Kim, D. W., Han, J. H., and Choe, J. (1997) *Eur. J. Biochem.* **250,** 47–54

136.   Preugschat, F., Averett, D. R., Clarke, B. E., and Porter, D. J. T. (1996) *J. Biol. Chem.* **271,** 24449–24457

137.   Levin, M. K., and Patel, S. S. (2002) *J. Biol. Chem.* **277,** 29377–29385

138.   Levin, M. K., Gurjar, M. M., and Patel, S. S. (2003) *J. Biol. Chem.* **278,** 23311–23316

139.   Ponomarev, M. A., Timofeev, V. P., and Levitsky, D. I. (1995) *FEBS Lett.* **371,** 261–263

140.   Porter, D. J. (1998) *J. Biol. Chem.* **273,** 7390–7396

141.   Kim, D. W., Gwack, Y., Han, J. H., and Choe, J. (1995) *Biochem. Biophys. Res. Commun.* **215,** 160–166

142.   Levin, M. K., and Patel, S. S. (1999) *J. Biol. Chem.* **274,** 31839–31846

143.   Velankar, S. S., Soultanas, P., Dillingham, M. S., Subramanya, H. S., and Wigley, D. B. (1999) *Cell* **97,** 75–84

144.   Korolev, S., Hsieh, J., Gauss, G. H., Lohman, T. M., and Waksman, G. (1997) *Cell* **90,** 635–647

145.   Du, M. X., Johnson, R. B., Sun, X. L., Staschke, K. A., Colacino, J., and Wang, Q. M. (2002) *Biochem. J.* **363,** 147–155

146.   Gibrat, J. F., Madej, T., and Bryant, S. H. (1996) *Curr. Opin. Struct. Biol.* **6,** 377–385

Lab Work and Data Analysis in this chapter, kindly contributed by Konstantina Dragoumani

Appendices

The Homology Modelling Data for the Helicase and the Polymerase Projects.

Table XX. The Yellow Fever Helicase Model's Procheck analysis summary.

```
+----------<<< P R O C H E C K    S U M M A R Y  >>>----------+
| YF   2.2                                          435 residues |
| Ramachandran plot:   81.5% core   13.4% allow    3.2% gener    1.9% disall |
| Gly & Pro Ramach:     5 labelled residues (out of  60)        |
| Chi1-chi2 plots:      1 labelled residues (out of 242)        |
| Main-chain params:    5 better     0 inside     1 worse       |
| Side-chain params:    5 better     0 inside     0 worse       |
| G-factors          Dihedrals:  -.17 Covalent:  -.23    Overall:   -.17 |
| M/c bond lengths: 83.6% within limits 16.4% highlighted       |
| M/c bond angles:  62.4% within limits 37.6% highlighted    8 off graph |
| Planar groups:    65.6% within limits 34.4% highlighted    9 off graph |
+----------------------------------------------------------------+
```

Table XX. The Japanese Encephalitis Helicase Model's Procheck analysis summary.

```
+----------<<< P R O C H E C K    S U M M A R Y  >>>----------+
| je   2.2                                          435 residues |
| Ramachandran plot:   79.9% core   15.1% allow    3.3% gener    1.6% disall |
| Gly & Pro Ramach:    11 labelled residues (out of  69)        |
| Chi1-chi2 plots:      2 labelled residues (out of 233)        |
| Main-chain params:    5 better     0 inside     1 worse       |
| Side-chain params:    5 better     0 inside     0 worse       |
| G-factors          Dihedrals:  -.16 Covalent:  -.23    Overall:   -.18 |
| M/c bond lengths: 69.2% within limits 30.8% highlighted    9 off graph |
| M/c bond angles:  67.1% within limits 32.9% highlighted    7 off graph |
| Planar groups:    70.0% within limits 30.0% highlighted   13 off graph |
+----------------------------------------------------------------+
```

Table XX. The Dengue Polymerase Model's Procheck analysis summary.

```
+----------<<< P R O C H E C K    S U M M A R Y  >>>----------+
| den   2.2                                         566 residues |
| Ramachandran plot:   79.7% core   16.9% allow    2.0% gener    1.4% disall |
| Gly & Pro Ramach:     4 labelled residues (out of  56)        |
| Chi1-chi2 plots:      0 labelled residues (out of 360)        |
| Main-chain params:    5 better     0 inside     1 worse       |
| Side-chain params:    5 better     0 inside     0 worse       |
| G-factors          Dihedrals:  -.04 Covalent:  -.22    Overall:   -.12 |
| M/c bond lengths: 63.5% within limits 36.5% highlighted   23 off graph |
| M/c bond angles:  65.0% within limits 35.0% highlighted   13 off graph |
| Planar groups:    65.6% within limits 34.4% highlighted   19 off graph |
+----------------------------------------------------------------+
```

Table XX. The West Nile Polymerase Model's Procheck analysis summary.

```
+----------<<< P R O C H E C K    S U M M A R Y  >>>----------+
| c:\dimitris\pdbs\wnv   2.0                        566 residues |
| Ramachandran plot:   68.2% core   24.6% allow    6.1% gener    1.0% disall |
| Gly & Pro Ramach:     4 labelled residues (out of  76)        |
| Chi1-chi2 plots:      8 labelled residues (out of 346)        |
| Main-chain params:    3 better     0 inside     3 worse       |
| Side-chain params:    5 better     0 inside     0 worse       |
| G-factors          Dihedrals:  -.73 Covalent:  -.86    Overall:   -.80 |
| M/c bond lengths: 85.9% within limits 14.1% highlighted       |
| M/c bond angles:  67.6% within limits 32.4% highlighted       |
| Planar groups:    94.7% within limits  5.3% highlighted       |
+----------------------------------------------------------------+
```

Table XX. The Japanese Encephalitis Polymerase Model's Procheck analysis summary

```
+----------<<< P R O C H E C K   S U M M A R Y >>>----------+
| je   2.2                                       566 residues |
| Ramachandran plot:   68.7% core   24.1% allow   6.4% gener   .8% disall |
| Gly & Pro Ramach:     6 labelled residues (out of  78)      |
| Chi1-chi2 plots:     11 labelled residues (out of 341)      |
| Main-chain params:    3 better     0 inside     3 worse     |
| Side-chain params:    5 better     0 inside     0 worse     |
| G-factors          Dihedrals:  -.73 Covalent:  -.87    Overall:  -.80 |
| M/c bond lengths: 88.1% within limits  11.9% highlighted    |
| M/c bond angles:  67.6% within limits  32.4% highlighted    |
| Planar groups:    92.0% within limits   8.0% highlighted    |
+------------------------------------------------------------+
```

Table XX. The Yellow Fever Polymerase Model's Procheck analysis summary.

```
+----------<<< P R O C H E C K   S U M M A R Y >>>----------+
| c:\dimitris\pdbs\yf   2.0                      566 residues |
| Ramachandran plot:   67.5% core   25.3% allow   6.6% gener   .6% disall |
| Gly & Pro Ramach:     5 labelled residues (out of  63)      |
| Chi1-chi2 plots:      4 labelled residues (out of 348)      |
| Main-chain params:    3 better     0 inside     3 worse     |
| Side-chain params:    5 better     0 inside     0 worse     |
| G-factors          Dihedrals:  -.73 Covalent:  -.88    Overall:  -.80 |
| M/c bond lengths: 86.8% within limits  13.2% highlighted    |
| M/c bond angles:  64.5% within limits  35.5% highlighted    1 off graph |
| Planar groups:    93.3% within limits   6.7% highlighted    |
+------------------------------------------------------------+
```

The Ramachandran plots of the Models of the Yellow Fever (left) and the Japanese Encephalitis (right) Helicases.

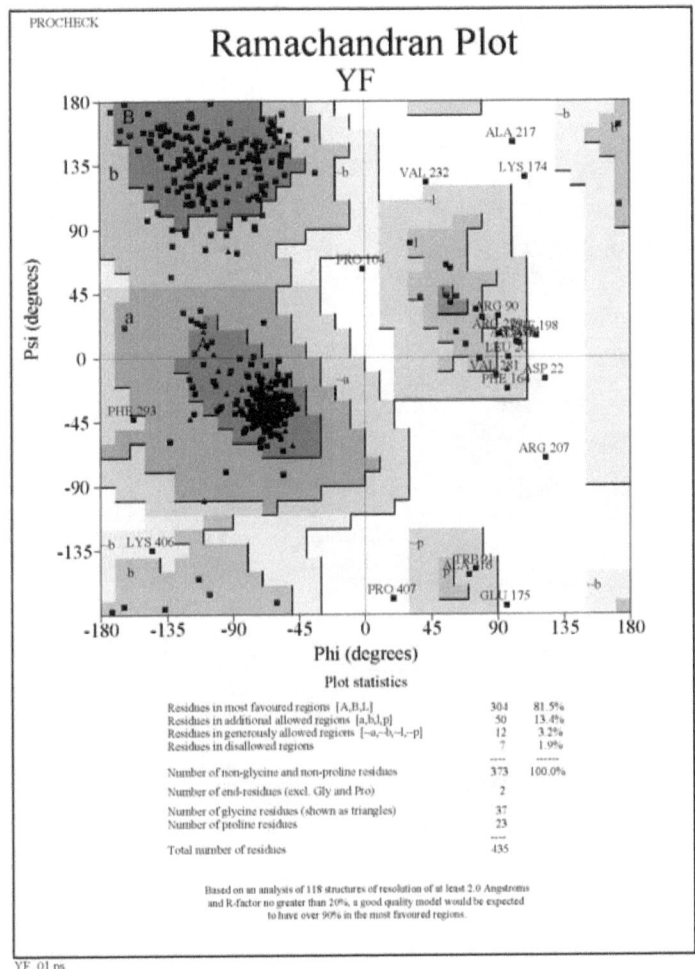

PROCHECK

Ramachandran Plot
YF

Plot statistics

Residues in most favoured regions [A,B,L]	304	81.5%
Residues in additional allowed regions [a,b,l,p]	50	13.4%
Residues in generously allowed regions [~a,~b,~l,~p]	12	3.2%
Residues in disallowed regions	7	1.9%
	----	------
Number of non-glycine and non-proline residues	373	100.0%
Number of end-residues (excl. Gly and Pro)	2	
Number of glycine residues (shown as triangles)	37	
Number of proline residues	23	

Total number of residues	435	

Based on an analysis of 118 structures of resolution of at least 2.0 Angstroms
and R-factor no greater than 20%, a good quality model would be expected
to have over 90% in the most favoured regions.

YF_01.ps

Ramachandran Plot
je

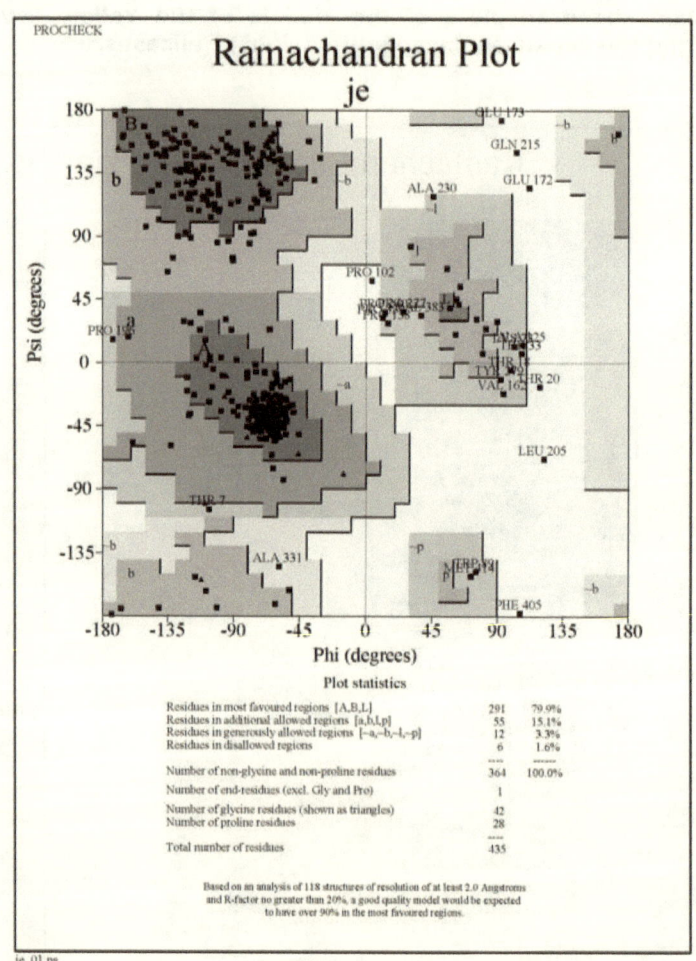

Psi (degrees)

Phi (degrees)

Plot statistics

Residues in most favoured regions [A,B,L]	291	79.9%
Residues in additional allowed regions [a,b,l,p]	55	15.1%
Residues in generously allowed regions [~a,~b,~l,~p]	12	3.3%
Residues in disallowed regions	6	1.6%
	----	----
Number of non-glycine and non-proline residues	364	100.0%
Number of end-residues (excl. Gly and Pro)	1	
Number of glycine residues (shown as triangles)	42	
Number of proline residues	28	

Total number of residues	435	

Based on an analysis of 118 structures of resolution of at least 2.0 Angstroms
and R-factor no greater than 20%, a good quality model would be expected
to have over 90% in the most favoured regions.

je_01.ps

The Ramachandran plots of the Models of the Dengue (top left), the Japanese Encephalitis (top right), the West Nile (bottom left) and the Yellow Fever (bottom right) Polymerases.

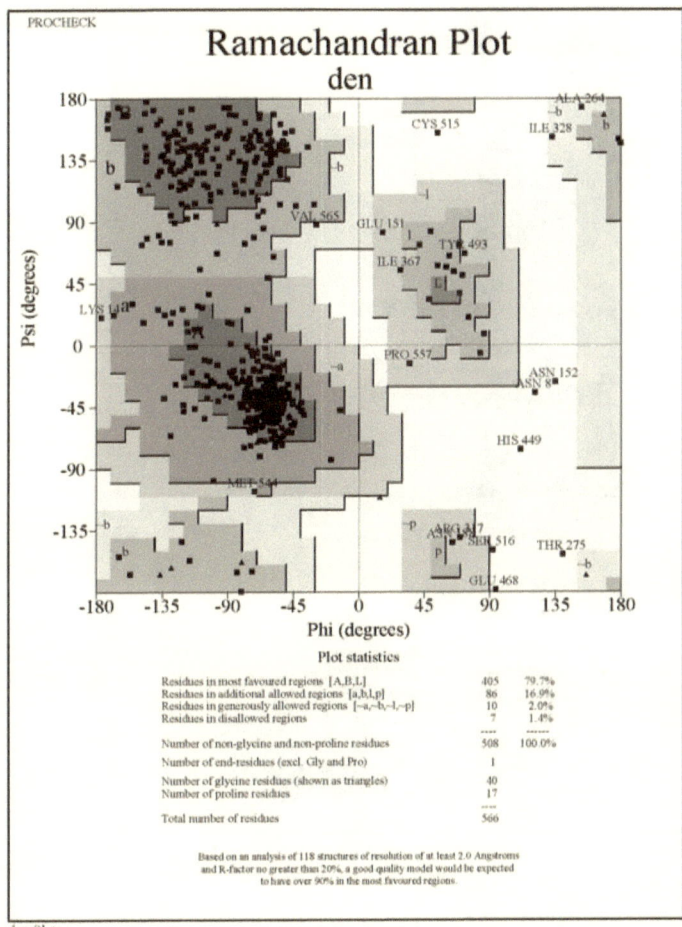

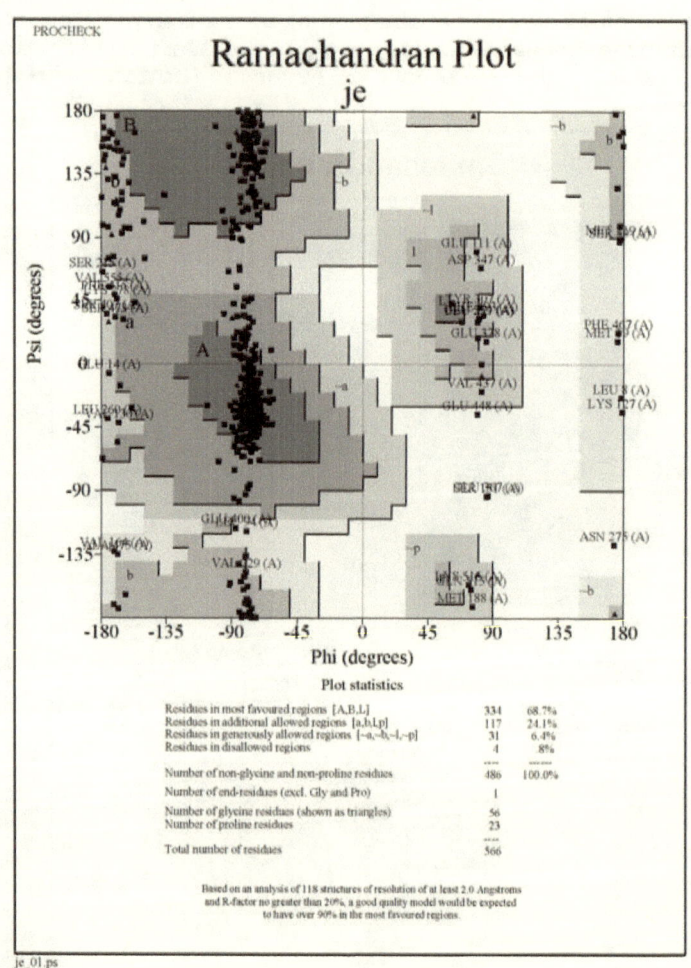

PROCHECK

Ramachandran Plot
je

Plot statistics

Residues in most favoured regions [A,B,L]	334	68.7%
Residues in additional allowed regions [a,b,l,p]	117	24.1%
Residues in generously allowed regions [~a,~b,~l,~p]	31	6.4%
Residues in disallowed regions	4	.8%
	----	------
Number of non-glycine and non-proline residues	486	100.0%
Number of end-residues (excl. Gly and Pro)	1	
Number of glycine residues (shown as triangles)	56	
Number of proline residues	23	

Total number of residues	566	

Based on an analysis of 118 structures of resolution of at least 2.0 Angstroms
and R-factor no greater than 20%, a good quality model would be expected
to have over 90% in the most favoured regions.

je_01.ps

188

Ramachandran Plot
c:dimitrispdbswny

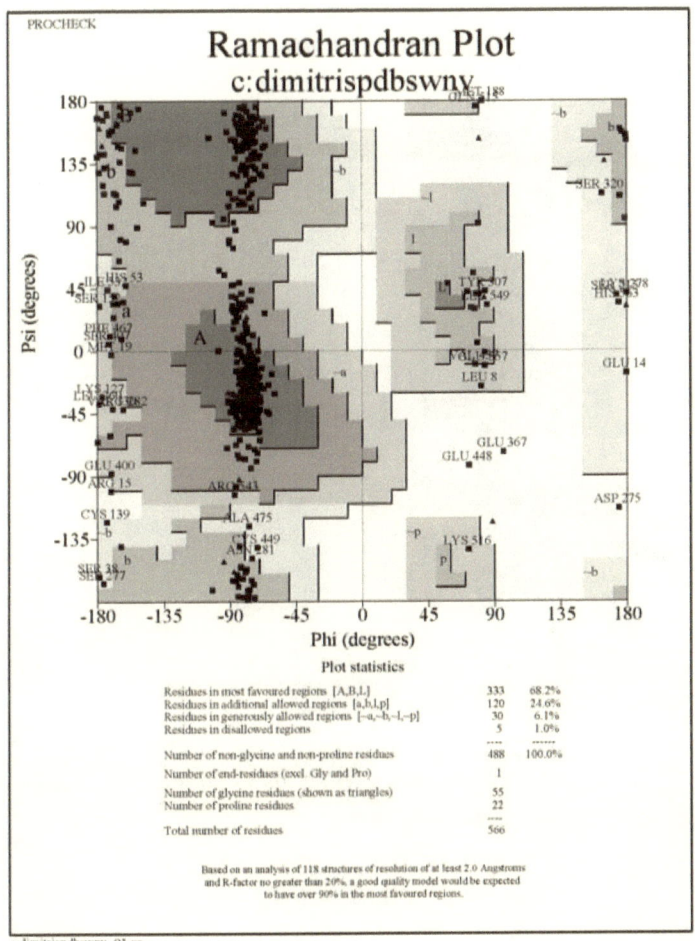

Psi (degrees) / **Phi (degrees)**

Plot statistics

Residues in most favoured regions [A,B,L]	333	68.2%
Residues in additional allowed regions [a,b,l,p]	120	24.6%
Residues in generously allowed regions [~a,~b,~l,~p]	30	6.1%
Residues in disallowed regions	5	1.0%
	----	------
Number of non-glycine and non-proline residues	488	100.0%
Number of end-residues (excl. Gly and Pro)	1	
Number of glycine residues (shown as triangles)	55	
Number of proline residues	22	

Total number of residues	566	

Based on an analysis of 118 structures of resolution of at least 2.0 Angstroms
and R-factor no greater than 20%, a good quality model would be expected
to have over 90% in the most favoured regions.

c:dimitrispdbswny_01.ps

189

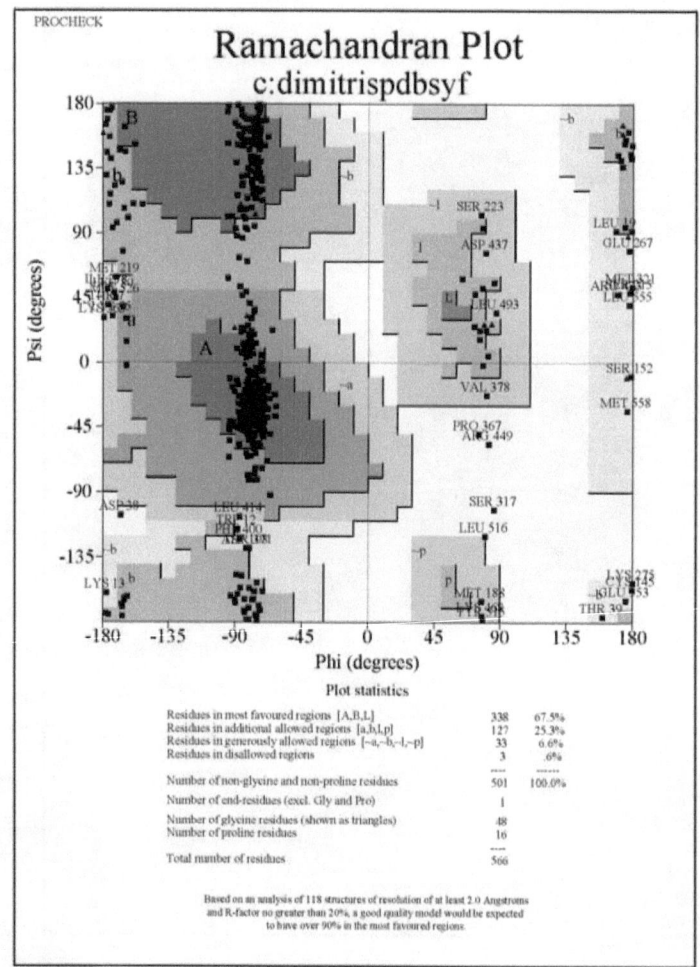

PROCHECK

Ramachandran Plot
c:dimitrispdbsyf

Phi (degrees)

Psi (degrees)

Plot statistics

Residues in most favoured regions [A,B,L]	338	67.5%
Residues in additional allowed regions [a,b,l,p]	127	25.3%
Residues in generously allowed regions [~a,~b,~l,~p]	33	6.6%
Residues in disallowed regions	3	.6%
	----	------
Number of non-glycine and non-proline residues	501	100.0%
Number of end-residues (excl. Gly and Pro)	1	
Number of glycine residues (shown as triangles)	48	
Number of proline residues	16	

Total number of residues	566	

Based on an analysis of 118 structures of resolution of at least 2.0 Angstroms
and R-factor no greater than 20%, a good quality model would be expected
to have over 90% in the most favoured regions.

c:dimitrispdbsyf_01.ps

190

Helicase – ssDNA/RNA contacts from LigPlot

Analytic Protein-DNA/RNA Contacts

The list of contacting atom pairs:

41 PRO (230) A	O <> 435 OURA (7) B	C5' D= 2.82 (B-S)	203 ARG (393) A	NH2 <> 434 OURA (6) B	C5 D= 3.52 (S-S)
43 VAL (232) A	CG2 <> 435 OURA (7) B	P D= 3.97 (S-B)	221 THR (411) A	CB <> 432 OURA (4) B	O3' D= 3.43 (S-B)
43 VAL (232) A	CG2 <> 435 OURA (7) B	C3' D= 3.57 (S-B)	221 THR (411) A	OG1 <> 433 OURA (5) B	P D= 3.46 (S-B)
43 VAL (232) A	CG2 <> 435 OURA (7) B	O5' D= 3.18 (S-S)	221 THR (411) A	OG1 <> 433 OURA (5) B	OP2 D= 2.54 (S-S)
43 VAL (232) A	CG2 <> 436 OURA (8) B	P D= 3.70 (S-B)	223 ALA (413) A	N <> 433 OURA (5) B	C5' D= 3.48 (B-S)
43 VAL (232) A	CG2 <> 436 OURA (8) B	OP1 D= 3.10 (S-B)	223 ALA (413) A	CA <> 433 OURA (5) B	C3' D= 3.56 (B-B)
66 GLY (255) A	N <> 436 OURA (8) B	P D= 3.60 (B-B)	223 ALA (413) A	CA <> 433 OURA (5) B	C5' D= 3.76 (B-S)
66 GLY (255) A	N <> 436 OURA (8) B	OP2 D= 3.06 (B-S)	223 ALA (413) A	CA <> 433 OURA (5) B	C4' D= 3.77 (B-S)
66 GLY (255) A	CA <> 436 OURA (8) B	P D= 3.93 (B-B)	223 ALA (413) A	CB <> 433 OURA (5) B	C3' D= 3.82 (S-B)
80 THR (269) A	CB <> 436 OURA (8) B	OP2 D= 3.28 (S-S)	223 ALA (413) A	CB <> 433 OURA (5) B	OP2 D= 3.03 (S-S)
80 THR (269) A	OG1 <> 436 OURA (8) B	OP2 D= 2.42 (S-S)	238 VAL (432) A	CG1 <> 431 OURA (3) B	C4' D= 3.73 (S-S)
82 GLY (271) A	CA <> 436 OURA (8) B	C5' D= 3.39 (B-S)	238 VAL (432) A	CG2 <> 431 OURA (3) B	C5' D= 3.55 (S-S)
82 GLY (271) A	C <> 436 OURA (8) B	C5' D= 3.00 (B-S)	238 VAL (432) A	CG2 <> 431 OURA (3) B	C4' D= 3.54 (S-S)
82 GLY (271) A	C <> 436 OURA (8) B	C4' D= 3.59 (B-S)	240 GLN (434) A	CG <> 432 OURA (4) B	C4 D= 3.82 (S-S)
83 LYS (272) A	N <> 436 OURA (8) B	C5' D= 2.94 (B-S)	240 GLN (434) A	CG <> 432 OURA (4) B	O4 D= 2.72 (S-S)
83 LYS (272) A	CA <> 436 OURA (8) B	C5' D= 3.52 (B-S)	240 GLN (434) A	CD <> 432 OURA (4) B	O4 D= 3.02 (S-S)
86 ALA (275) A	CB <> 436 OURA (8) B	O3' D= 3.34 (S-B)	240 GLN (434) A	NE2 <> 432 OURA (4) B	O4 D= 2.89 (S-S)
179 HIS (369) A	CD2 <> 431 OURA (3) B	C5' D= 3.64 (S-S)	254 THR (448) A	CG2 <> 432 OURA (4) B	C5 D= 3.75 (S-S)
179 HIS (369) A	NE2 <> 431 OURA (3) B	P D= 3.61 (S-B)	254 THR (448) A	CG2 <> 432 OURA (4) B	C4 D= 3.83 (S-S)
179 HIS (369) A	NE2 <> 431 OURA (3) B	OP2 D= 2.76 (S-S)	254 THR (448) A	CG2 <> 432 OURA (4) B	O4 D= 3.35 (S-S)
179 HIS (369) A	NE2 <> 431 OURA (3) B	C5' D= 3.73 (S-S)	256 THR (450) A	OG1 <> 430 OURA (2) B	O3' D= 2.95 (S-B)
179 HIS (369) A	O <> 432 OURA (4) B	C5' D= 2.96 (B-S)	306 TRP (501) A	CG <> 436 OURA (8) B	C2 D= 3.81 (S-S)
179 HIS (369) A	O <> 432 OURA (4) B	C4' D= 3.37 (B-S)	306 TRP (501) A	CD2 <> 436 OURA (8) B	N1 D= 3.67 (S-S)
180 SER (370) A	CA <> 432 OURA (4) B	OP2 D= 3.17 (B-S)	306 TRP (501) A	CD2 <> 436 OURA (8) B	C2 D= 3.53 (S-S)
180 SER (370) A	CA <> 432 OURA (4) B	C5' D= 3.67 (B-S)	306 TRP (501) A	NE1 <> 436 OURA (8) B	N3 D= 3.45 (S-S)
180 SER (370) A	CB <> 432 OURA (4) B	OP2 D= 3.12 (S-S)	306 TRP (501) A	CE2 <> 436 OURA (8) B	C4 D= 3.55 (S-S)
181 LYS (371) A	N <> 432 OURA (4) B	OP2 D= 2.93 (B-S)	306 TRP (501) A	CE2 <> 436 OURA (8) B	N3 D= 3.33 (S-S)
203 ARG (393) A	N <> 433 OURA (5) B	P D= 3.90 (B-S)	306 TRP (501) A	CE2 <> 436 OURA (8) B	C2 D= 3.55 (S-S)
203 ARG (393) A	N <> 433 OURA (5) B	OP2 D= 2.80 (B-S)	306 TRP (501) A	CE3 <> 436 OURA (8) B	C1' D= 3.73 (S-S)
203 ARG (393) A	CB <> 433 OURA (5) B	P D= 3.68 (S-B)	306 TRP (501) A	CE3 <> 436 OURA (8) B	N1 D= 3.59 (S-S)
203 ARG (393) A	CB <> 433 OURA (5) B	OP1 D= 3.36 (S-B)	306 TRP (501) A	CE3 <> 436 OURA (8) B	C6 D= 3.85 (S-S)
203 ARG (393) A	CB <> 433 OURA (5) B	OP2 D= 3.41 (S-S)	306 TRP (501) A	CZ2 <> 436 OURA (8) B	C6 D= 3.75 (S-S)
203 ARG (393) A	NE <> 433 OURA (5) B	C2' D= 3.08 (S-S)	306 TRP (501) A	CZ2 <> 436 OURA (8) B	C5 D= 3.35 (S-S)
203 ARG (393) A	NE <> 434 OURA (6) B	OP1 D= 3.06 (S-B)	306 TRP (501) A	CZ2 <> 436 OURA (8) B	C4 D= 3.21 (S-S)
203 ARG (393) A	NE <> 434 OURA (6) B	C5 D= 3.68 (S-S)	306 TRP (501) A	CZ2 <> 436 OURA (8) B	N3 D= 3.54 (S-S)
203 ARG (393) A	CZ <> 434 OURA (6) B	OP1 D= 3.30 (S-S)	306 TRP (501) A	CZ3 <> 436 OURA (8) B	N1 D= 3.75 (S-S)
203 ARG (393) A	CZ <> 434 OURA (6) B	C6 D= 3.47 (S-S)	306 TRP (501) A	CZ3 <> 436 OURA (8) B	C6 D= 3.48 (S-S)
203 ARG (393) A	CZ <> 434 OURA (6) B	C5 D= 3.48 (S-S)	306 TRP (501) A	CH2 <> 436 OURA (8) B	C6 D= 3.44 (S-S)
203 ARG (393) A	NH2 <> 434 OURA (6) B	P D= 3.71 (S-B)	306 TRP (501) A	CH2 <> 436 OURA (8) B	C5 D= 3.29 (S-S)
203 ARG (393) A	NH2 <> 434 OURA (6) B	OP1 D= 2.92 (S-B)	306 TRP (501) A	CH2 <> 436 OURA (8) B	C4 D= 3.72 (S-S)
203 ARG (393) A	NH2 <> 434 OURA (6) B	O5' D= 3.28 (S-S)	307 TYR (502) A	OH <> 436 OURA (8) B	O2 D= 2.93 (S-S)
203 ARG (393) A	NH2 <> 434 OURA (6) B	C2' D= 3.50 (S-S)	361 ASN (556) A	OD1 <> 433 OURA (5) B	O4 D= 2.83 (S-S)
203 ARG (393) A	NH2 <> 434 OURA (6) B	C6 D= 3.00 (S-S)		

Helicase – Cysteine (+small molecule attached) contacts from LigPlot

Analytic Protein - Cofactor Contacts

The list of contacting atom pairs:

Protein Atom:				Ligand Atom:				
237 THR	(430) A	N	<>	438 CME	(431) A	N	D= 3.63	(B-D)
237 THR	(430) A	CA	<>	438 CME	(431) A	CA	D= 3.80	(B-D)
237 THR	(430) A	C	<>	438 CME	(431) A	CB	D= 3.22	(B-D)
237 THR	(430) A	C	<>	438 CME	(431) A	C	D= 3.63	(B-D)
237 THR	(430) A	O	<>	438 CME	(431) A	CA	D= 2.72	(B-D)
237 THR	(430) A	O	<>	438 CME	(431) A	CB	D= 3.35	(B-D)
237 THR	(430) A	CB	<>	438 CME	(431) A	N	D= 3.12	(S-D)
238 VAL	(432) A	N	<>	438 CME	(431) A	N	D= 3.57	(B-D)
238 VAL	(432) A	N	<>	438 CME	(431) A	CB	D= 3.24	(B-D)
238 VAL	(432) A	CA	<>	438 CME	(431) A	CA	D= 3.79	(B-D)
238 VAL	(432) A	CA	<>	438 CME	(431) A	O	D= 2.79	(B-D)
238 VAL	(432) A	C	<>	438 CME	(431) A	C	D= 3.17	(B-D)
238 VAL	(432) A	C	<>	438 CME	(431) A	O	D= 3.43	(B-D)
238 VAL	(432) A	CB	<>	438 CME	(431) A	C	D= 3.69	(S-D)
256 THR	(450) A	CA	<>	438 CME	(431) A	O	D= 3.37	(B-D)
257 LEU	(451) A	N	<>	438 CME	(431) A	O	D= 2.81	(B-D)
257 LEU	(451) A	O	<>	438 CME	(431) A	N	D= 2.74	(B-D)
259 GLN	(453) A	CA	<>	438 CME	(431) A	SG	D= 3.87	(B-D)
259 GLN	(453) A	CA	<>	438 CME	(431) A	SD	D= 3.74	(B-D)
260 ASP	(454) A	N	<>	438 CME	(431) A	SD	D= 3.88	(B-D)
260 ASP	(454) A	OD1	<>	438 CME	(431) A	SD	D= 3.62	(S-D)
263 SER	(457) A	CB	<>	438 CME	(431) A	SD	D= 3.50	(S-D)

Interaction Maps of the Helicase from LigPlot

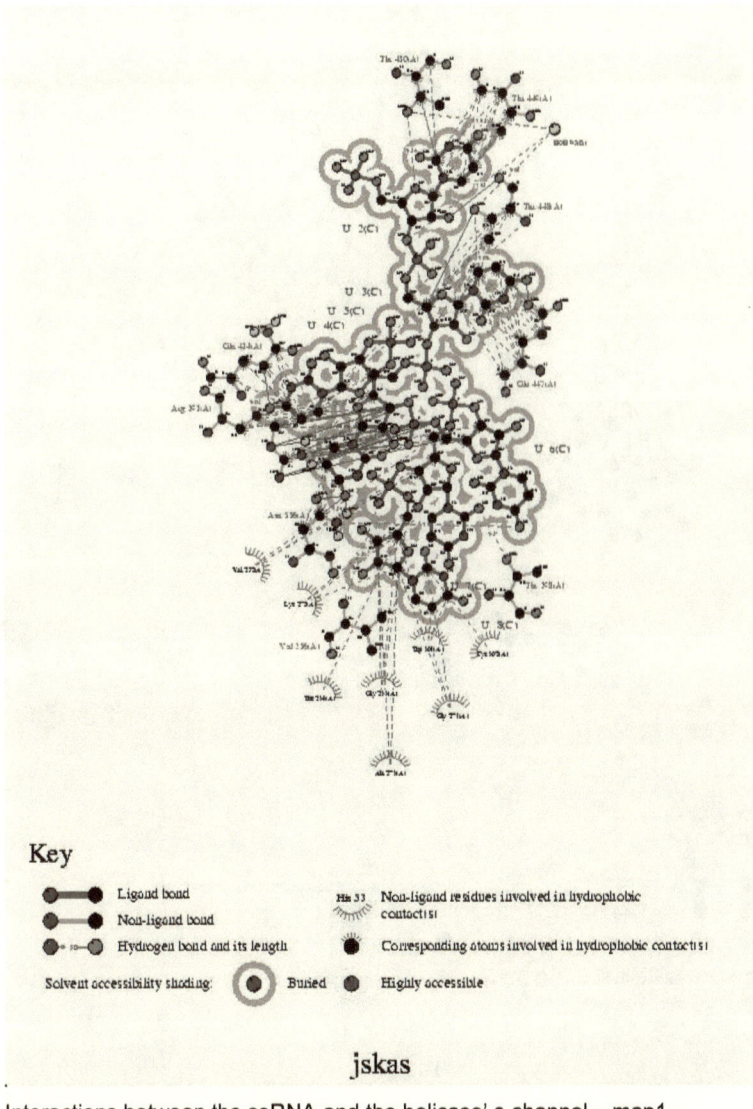

Key

●━━━●	Ligand bond
●──●	Non-ligand bond
●··30··●	Hydrogen bond and its length
Solvent accessibility shading: ◉ Buried ● Highly accessible	

His 33   Non-ligand residues involved in hydrophobic contact(s)

●   Corresponding atoms involved in hydrophobic contact(s)

jskas

Interactions between the ssRNA and the helicase' s channel – map1

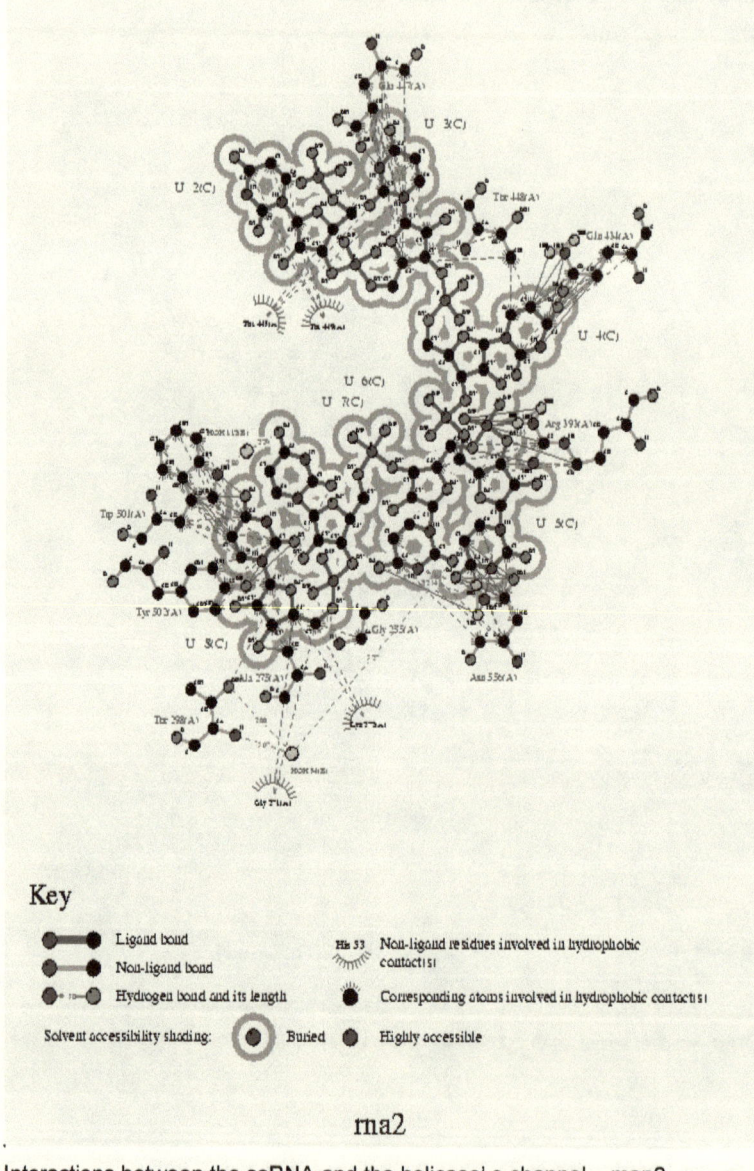

Key

Ligand bond
Non-ligand bond
Hydrogen bond and its length

His 53 Non-ligand residues involved in hydrophobic contact(s)

Corresponding atoms involved in hydrophobic contact(s)

Solvent accessibility shading: Buried Highly accessible

rna2

Interactions between the ssRNA and the helicase' s channel – map2

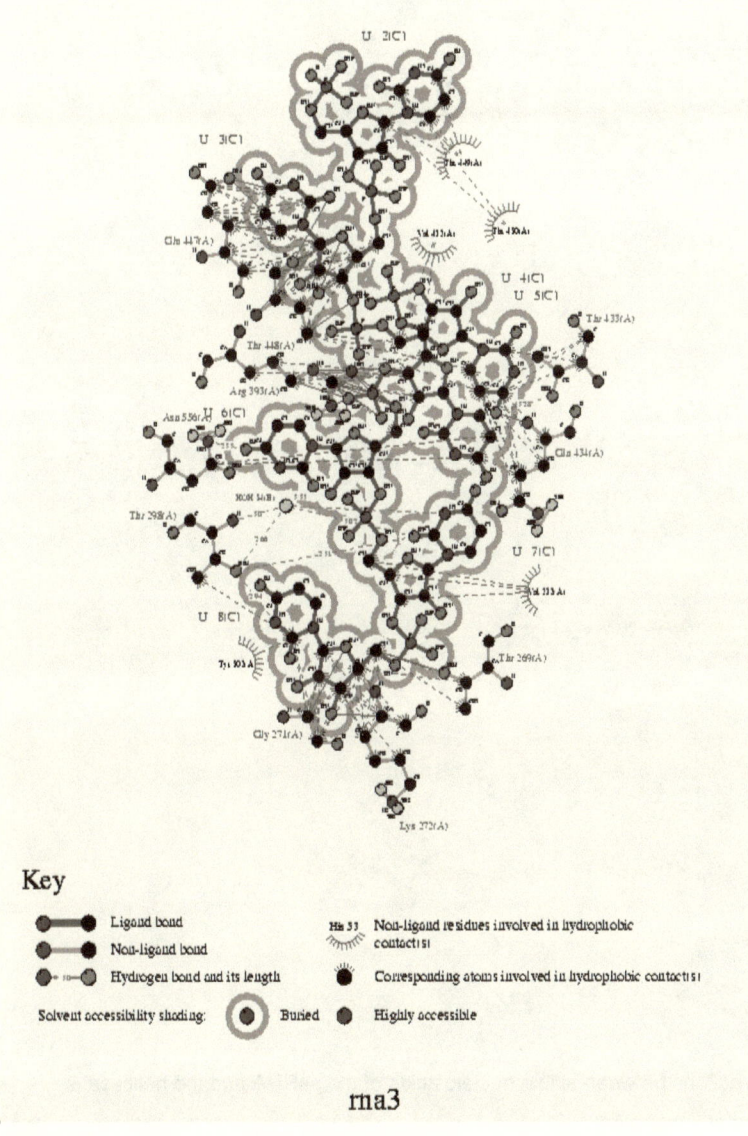

rna3

Interactions between the ssRNA and the helicase' s channel – map3

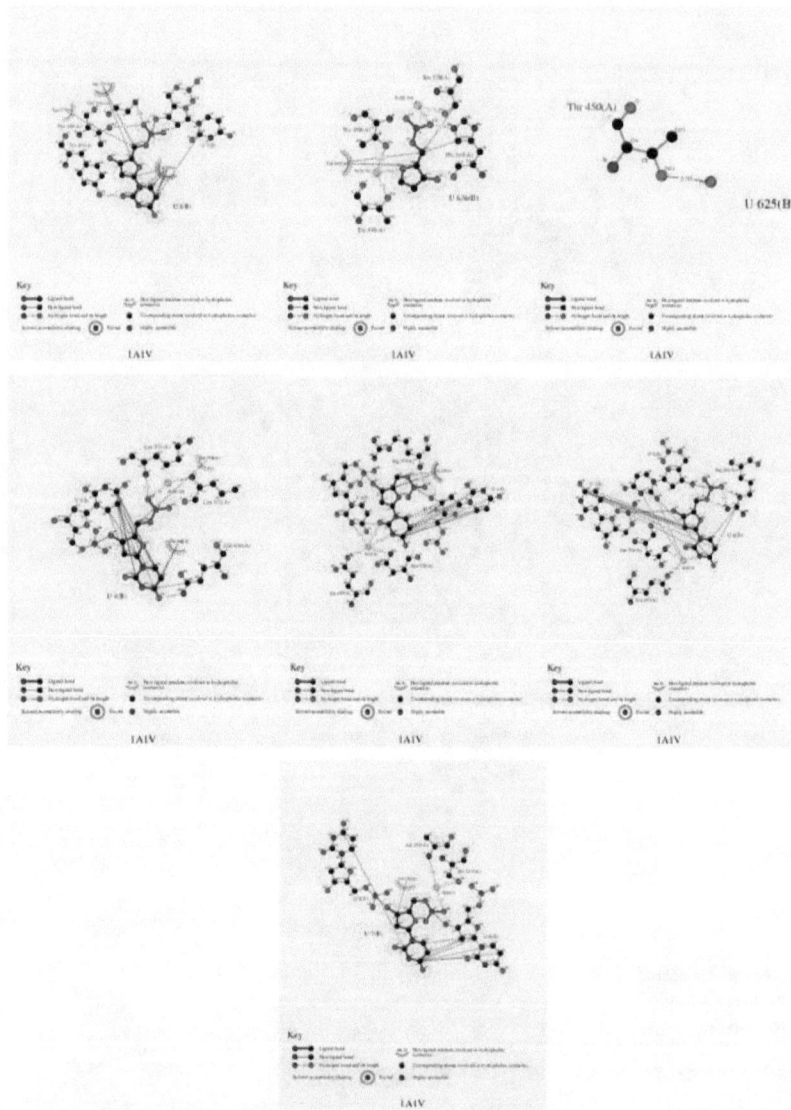

Interactions between single nucleic acids of the ssRNA and the helicase's channel – maps 4-10

LPC analysis of the Cysteine (+ attached compound)

Key to LPC Tables.

Dist	nearest distance (Å) between atoms of the ligand and the residue
Surf	contact surface area (Å2) between the ligand and the residue
HB	hydrophilic-hydrophilic contact (hydrogen bond)
Arom	aromatic-aromatic contact
Phob	hydrophobic-hydrophobic contact
DC	hydrophobic-hydrophilic contact (destabilizing contact)
+/-	indicates presence/absence of a specific contacts
*	indicates residues contacting ligand by their side chain (including CA atoms)

Residues in contact with ligand CME 431A in PDB entry 1A1V

Residue		Dist	Surf	HB	Arom	Phob	DC
369A	HIS*	4.5	4.7	-	-	-	-
412A	ASP*	3.9	16.2	+	-	-	+
430A	THR*	1.3	76.7	-	-	-	+
432A	VAL*	1.3	76.1	+	-	-	+
433A	THR*	3.9	15.8	+	-	-	+
449A	THR	3.5	0.9	-	-	-	+
450A	THR*	3.4	3.0	-	-	-	-
451A	LEU*	2.7	44.7	+	-	-	+
452A	PRO	4.1	3.4	-	-	-	-
453A	GLN*	3.7	13.2	-	-	-	-
454A	ASP*	3.6	17.9	-	-	-	+
457A	SER*	3.5	29.6	-	-	-	+
461A	ARG*	4.2	10.5	-	-	-	+
481A	ARG*	4.1	11.7	-	-	-	+
4B	U*	5.7	2.7	-	-	-	-

Residues in contact with ligand SO41001 in PDB entry 1A1V

Residue		Dist	Surf	HB	Arom	Phob	DC
206A	THR*	3.3	16.1	-	-	-	+
207A	GLY*	2.9	28.5	+	-	-	-
208A	SER*	3.0	11.4	+	-	-	-
209A	GLY*	2.8	36.1	+	-	-	-
210A	LYS*	2.9	44.1	+	-	-	+
211A	SER*	2.9	28.7	+	-	-	-
212A	THR*	4.7	2.8	+	-	-	-
290A	ASP*	4.8	1.2	+	-	-	-

```
   291A   GLU*    5.7     1.4     +       -           -       -
```

Residues in contact with ligand ADC 435 in PDB entry 1A7A

```
                                        Specific contacts
                                        ---------------------------
        Residue     Dist    Surf    HB      Arom    Phob    DC
--------------------------------------------------------------------
     433    NAH     3.3     39.0    -       -       +       +
      54A   LEU*    3.6     19.4    -       -       +       +
      55A   HIS*    3.4     40.8    -       +       +       +
      57A   THR*    2.5     24.0    +       -       +       -
      59A   GLU*    2.7     31.5    +       -       -       +
      60A   THR*    3.3     20.8    +       -       -       -
     131A   ASP*    3.8     20.1    +       -       -       +
     156A   GLU*    2.6     27.2    +       -       -       +
     157A   THR*    3.0     28.2    +       -       +       +
     186A   LYS*    3.2     17.8    +       -       -       -
     190A   ASP*    2.7     24.9    +       -       -       -
     301A   HIS*    5.2      2.7    -       -       +       +
     344A   LEU*    5.0      2.7    -       -       +       -
     346A   ASN*    4.4      0.7    -       -       -       -
     347A   LEU*    3.7     19.1    -       -       +       +
     351A   MSE     3.4     17.4    +       -       -       -
     353A   HIS*    2.8     58.1    +       +       -       +
     358A   MSE     2.8     21.4    -       -       +       +
     362A   PHE*    3.7      9.6    -       +       -       -
--------------------------------------------------------------------
```

Residues in contact with ligand NAH 433 in PDB entry 1A7A

```
                                        Specific contacts
                                        ---------------------------
        Residue     Dist    Surf    HB      Arom    Phob    DC
--------------------------------------------------------------------
     435    ADC     3.3     45.3    -       -       +       -
     157A   THR*    2.7     26.5    +       -       -       -
     158A   THR*    2.6     58.9    +       -       -       +
     159A   THR*    2.7     34.4    +       -       -       -
     186A   LYS*    4.1      0.2    -       -       -       -
     190A   ASP*    3.7     11.9    -       -       -       -
     191A   ASN*    3.5     31.0    +       -       -       -
     195A   CYS*    4.3      2.7    -       -       -       -
     219A   ALA*    4.7      1.6    -       -       +       -
     220A   GLY*    3.5     14.5    -       -       -       -
     222A   GLY*    2.7     47.5    -       -       -       -
     223A   ASP*    3.1     10.7    +       -       -       +
     224A   VAL*    2.7     58.2    +       -       +       +
     242A   THR*    3.8     23.1    -       -       -       -
     243A   GLU*    2.8     57.1    +       -       -       +
     244A   ILE*    3.4     53.3    -       -       +       +
     245A   ASP*    3.4      6.4    -       -       -       +
     248A   ASN*    3.4     11.8    +       -       -       -
     275A   THR     3.2     39.3    -       -       -       +
     276A   THR*    3.2     52.7    +       -       +       +
     277A   GLY*    4.1     13.7    -       -       -       -
```

```
278A   CYS*   3.4   18.4    -     -     -     -
281A   ILE*   3.1   31.3    -     -     +     +
299A   ILE*   3.0   35.2    +     -     +     +
300A   GLY*   3.5    8.9    -     -     -     -
301A   HIS*   3.0   32.0    +     -     -     +
305A   GLU*   4.9    0.2    +     -     -     -
344A   LEU*   3.9   11.1    -     -     -     +
346A   ASN*   2.7   34.1    +     -     -     +
353A   HIS*   3.1   21.3    +     +     -     -
407B   THR*   4.5    9.9    -     -     +     +
409B   LEU*   4.1    1.2    -     -     -     +
413B   GLN*   3.6   16.8    +     -     -     -
417B   LEU*   3.9    9.2    -     -     +     +
419B   MSE    5.0    1.6    -     -     +     -
425B   PHE*   5.3    0.2    -     -     -     +
426B   LYS*   2.7   45.8    +     -     -     +
430B   TYR*   2.9   27.4    +     -     -     -
-----------------------------------------------------------
```

The Dengue & West Nile Virus Helicase Sequences

```
LOCUS       NP_739587                   615 aa     linear   VRL 29-APR-2003
DEFINITION  NS3 protein; protease; RNA-helicase; ATPase; component of
capping
            enzyme (RNA thriphosphatase) [Dengue virus].
ACCESSION   NP_739587
VERSION     NP_739587.1  GI:25059134
DBSOURCE    REFSEQ: accession NP_056776.1
KEYWORDS    .
SOURCE      Dengue virus
ORGANISM    Dengue virus
            Viruses; ssRNA positive-strand viruses, no DNA stage;
            Flaviviridae;
            Flavivirus; Dengue virus group.
REFERENCE   1  (residues 1 to 615)
AUTHORS     Hahn,Y.S., Galler,R., Hunkapiller,T., Dalrymple,J.M.,
            Strauss,J.H.
            and Strauss,E.G.
   TITLE    Nucleotide sequence of dengue 2 RNA and comparison of the
            encoded
            proteins with those of other flaviviruses
JOURNAL     Virology 162 (1), 167-180 (1988)
MEDLINE     88101365
PUBMED      2827375
COMMENT     PROVISIONAL REFSEQ: This record has not yet been subject to
            final NCBI review. The reference sequence was derived from
            NS3 protein. Draft entry and computer-readable sequence for
            [1] kindly provided by Y.S.Hahn, 19-APR-1988. The mature
            peptides were revised by the
            NCBI staff.
Method: conceptual translation.
FEATURES             Location/Qualifiers
     source          1..615
                     /organism="Dengue virus"
                     /db_xref="taxon:12637"
                     /note="type: 2"
     mat_peptide     1..615
                     /product="NS3 protein"
                     /note="protease; RNA-helicase; ATPase; component
                     of capping enzyme (RNA thriphosphatase)"
     Precursor       1..615
                     /derived_from="NP_056776.1:1476..2090"
ORIGIN
1    agvlwdvpsp ppvgkaeled gayrikqkgi lgysqigagv ykegtfhtmw hvtrgavlmh
61   kgkriepswa dvkkdlisyg ggwklegewk egeevqvlal epgknpravq tkpglfrtnt
121  gtigavsldf spgtsgspiv dkkgkvvgly gngvvtrsga yvsaiaqtek siednpeied
181  difrkrrlti mdlhpgagkt krylpaivre aikrglrtli laptrvvaae meealrglpi
241  ryqtpairae htgreivdlm chatftmrll spirvpnynl iimdeahftd pasiaargyi
301  strvemgeaa gifmtatppg srdpfpqsna pimdeereip erswnsghew vtdfkgktvw
361  fvpsiktgnd iaaclrkngk rviqlsrktf dseyvktrtn dwdfvvttdi semganfkae
421  rvidprrcmk pviltdgeer vilagpmpvt hssaaqrrgr igrnprnend qyiymgeple
481  ndedcahwke akmlldnint pegiipsmfe perekvdaid geyrlrgear ktfvdlmrrg
541  dlpvwlaykv aaeginyadr rwcfdgtrnn qileenveve iwtkegerkk lkprwldari
601  ysdplalkef aagrk
```

```
LOCUS       AAF18443                 619 aa     linear   VRL 16-DEC-1999
DEFINITION  NS3 polyprotein [West Nile virus].
ACCESSION   AAF18443 REGION: 1506..2124
VERSION     AAF18443.1  GI:6581070
DBSOURCE    locus AF202541 accession AF202541.1
SOURCE      West Nile virus
ORGANISM    West Nile virus
            Viruses; ssRNA positive-strand viruses, no DNA stage;
            Flaviviridae;
            Flavivirus; Japanese encephalitis virus group.
REFERENCE   1  (residues 1 to 619)
AUTHORS     Jia,X.Y., Briese,T., Jordan,I., Rambaut,A., Chi,H.C.,
            Mackenzie,J.S., Hall,R.A., Scherret,J. and Lipkin,W.I.
TITLE       Genetic analysis of West Nile New York 1999 encephalitis
            virus
JOURNAL     Lancet 354 (9194), 1971-1972 (1999)
MEDLINE     20086017
PUBMED      10622305
REFERENCE   2  (residues 1 to 619)
AUTHORS     Jia,X.Y., Briese,T., Jordan,I. and Lipkin,W.I.
TITLE       Direct Submission
JOURNAL     Submitted (06-NOV-1999) Emerging Diseases Laboratory, Dept.
            Microbiology & Molecular Genetics and Neurology, University
            Of California, Irvine, 3101 Gillespie Neuroscience
            Facility, Irvine, CA 92697-4292, USA

COMMENT     Method: conceptual translation.
FEATURES             Location/Qualifiers
     source          1..619
                     /organism="West Nile virus"
                     /strain="HNY1999"
                     /db_xref="taxon:11082"
                     /country="USA: New York"
                     /note="isolated from total brain RNA (patient
NYC99002) by
                     RT-PCR amplification"
     Protein         <1..>619
                     /product="polyprotein"
     mat_peptide     1..619
                     /product="non-structural protein NS3"
                     /name="putative"
     CDS             <1..>619
                     /coded_by="AF202541.1:55..10356"
ORIGIN
        1 ggvlwdtpsp keykkgdttt gvyrimtrgl lgsyqagagv mvegvfhtlw httkgaalms
       61 gegrldpywg svkedrlcyg gpwklqhkwn gqdevqmivv epgknvknvq tkpgvfktpe
      121 geigavtldf ptgtsgspiv dkngdvigly gngvimpngs yisaivqger mdepipagfe
      181 pemlrkkqit vldlhpgagk trrilpqiik eainrrlrta vlaptrvvaa emaealrglp
      241 iryqtsavpr ehngneivdv mchatlthrl msphrvpnyn lfvmdeahft dpasiaargy
      301 istkvelgea aaifmtatpp gtsdpfpesn spisdlqtei pdrawnsgye witeytgktv
      361 wfvpsvkmgn eialclqrag kkvvqlnrks yeteypkckn ddwdfvittd isemganfka
      421 srvidsrksv kptiitegeg rvilgepsav taasaaqrrg rigrnpsqvg deycygghtn
      481 eddsnfahwt earimldnin mpngliaqfy qperekvytm dgeyrlrgee rknflellrt
      541 adlpvwlayk vaaagvsyhd rrwcfdgprt ntilednnev evitklgerk ilrprwidar
      601 vysdhqalka fkdfasgkr
//
```

Tables of the modification of the ;ead compound for *in silico* testing on Hepc, Dengue and WNV Helicases.

The symbols used stand for:

√ → picks up high interaction with the protein (H bonds) and hydrophobic interaction too. It has a good conformation on the protein as well.

~√ →Does not H-bond but it establishes hydrophobic interaction and has a good docking conformation.

X →Was impossible to dock the compound in a successful way. No interaction established + the conformation of the compound is bad.

STRUCTURE	Number	HepC	Dengue	WNV
	1	X	X	X
	2	X	X	X
	3	X	X	X
	4	X	X	X
	5	X	X	X
	6	X	X	X
	7	X	X	X
	8	X	X	X
	9	X	X	X
	10	X	X	X

STRUCTURE	Number	HepC	Dengue	WNV
	11	X	X	X
	12	X	X	X
	13	X	X	X
	14	√	X	X
	15	X	X	X
	16	~√	X	X
	17	~√	~√	X
	18	X	X	X
	19	X	X	X
	20	~√	X	X

STRUCTURE	Number	HepC	Dengue	WNV
	21	X	X	X
	22	X	X	X
	23	X	X	X
	24	X	X	X
	25	~√	X	X
	26	X	X	X
	27	X	X	X
	28	X	X	X
	29	~√	√	√
	30	X	X	X
	31	√	√	~√

STRUCTURE	Number	HepC	Dengue	WNV
	32	√	√	~√
	33	X	X	X
	34	X	X	X
	35	~√	~√	~√
	36	X	X	X
	37	√	√	√
	38	√	√	√
	39	√	√	√
	40	√	√	√
	41	√	√	√
	42	√	√	√

STRUCTURE	Number	HepC	Dengue	WNV
	43	√	√	√
	44	√	√	√
	45	X	X	X
	46	√	√	√
	47	√	√	~√
	48	X	X	X
	48	X	X	X
	49	X	X	X
	50	X	X	X
	51	X	X	X

CHEMICAL STRUCTURE	No.	IC$_{50}$	HepC	Dengue	WNV
	1	10	√	~√	~√
	2	0.7	√	~√	~√
	3	0.7	~√	~√	~√
	4	0.7	√	√	√
	5	0.7	√	√	√
	6	0.7	√	√	√
	7	0.7	√	√	√

Cmpd#	Coupled NS3 ATPase (%inhibition@ 30 µM)	HPLC ATPase IC50 (µM)	Helicase Unwinding IC50 (µM)
1	B	F	F
2		G	
3		F	
4		F	H
5		E	G
6	C	G	
7	C	H	H
8	D		
9	C	G	
10	A	E	
11	B	E	
12	A	E	
13	A	E	E
14	D		
15	D		
16	A	E	
17	A	E	
18	A	E	

19	B	E	
20	A	F	
21	A	E	
22	A	E	
23		E	
24		E	
25		F	
26	A	E	
27		E	
28		E	
29		E	
30	A	E	
31	A	E	
32	A	E	
33	C	E	
34	B	F	
35	A	E	
36		E	
37		E	
38		E	
39		F	
40		F	
41		E	
42		H	
43		E	
44		E	
45		E	
46		E	
47		E	
48		E	
49		E	
50		H	
51		E	
52		G	
53		F	
54		E	
55			
56		H	

the Vertex compounds.

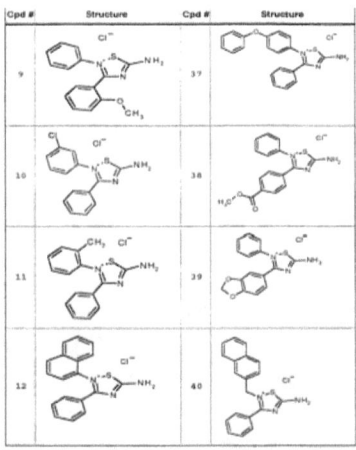

Cpd #	Structure	Cpd #	Structure
17		45	
18		46	
19		47	
20		48	

Cpd #	Structure	Cpd #	Structure
25		53	
26		54	
27		55	
28		56	

Cpd #	Structure	Cpd #	Structure
21		49	
22		50	
23		51	
24		52	

COMPLETE INHIBITION OF HEPATITIS C VIRUS HELICASE ACTIVITY BY HUMAN ANTIBODY FRAGMENTS CLONED BY PHAGE DISPLAY
Artsaenko, K. Tessmann, A. Erhardt, D. Haussinger, T. Heintges. *Dept of Gastroenterology, Hepatology and Infectious Diseases, Heinrich-Heine-University, Dusseldor$ Germany*

Sequence Alignments for the Homology Modelling Experiments.

```
1alvA                -HepC Helicase

Length = 429


 Score = 46.5 bits (108), Expect = 5e-05

 Identities = 71/272 (26 Positives = 104/272 (38 Gaps = 20/272 (7


Query: 196 GAGKTKRYLPAIVREAIKRGLRTLILAPTRVVAAEMEEALRGLPIRYQTPAIRAVHTGRE 255
           G GK     A    K     A T    A M A G    T  R  TG
Sbjct: 18  GSGKSTKVPAAYAAQGYKVLVLNPSVAATLGFGAYMSKA-HGVDPNIRT-GVRTITTGSP 75


Query: 256 IVDLMCHATFTMRLLSPVRVPNYNLIIMDEAHFTDPASIAARGYISTRVEMGEAAG---- 311
           I      T  L      Y II DE H TD  SI    I T    E AG
Sbjct: 76  IT----YSTYGKFLADGGSGGAYDIIICDECHSTDATSILG---IGTVLDQAETAGARLV 128


Query: 312 IFMTATPPGSRDPFPQSNAXXXXXXXXXXXXXSWNSGHEWVTDFKGKTVWFVPSIKAGNDI 371
                TATPPGS    P N                    G   F S K
Sbjct: 129 VLATATPPGS-VTVPHPNIEEVALSTTGEIPFYGKAIPLEVIKGGRHLIFCHSKKKCDEL 187


Query: 372 AACLRKNGKKVIQLSRKTFDSEYVKTRTNDWDFVVTTDISEMGANFKAERVIDPRRCMKP 431
           AA L  G      R  D     T    D VV       F    VID
Sbjct: 188 AAKLVALGINAVAYYR-GLDVSVIPT---SGDVVVVATDALFTGDF--DSVIDCNTVTQT 241


Query: 432 VILTDGEERVILAGPMPVTHSSAAQRRGRIGR 463
           V     I     P   S  QRRGR GR
Sbjct: 242 VDFSLDPTFTIETTTLPQDAVSRTQRRGRTGR 273
```

211

```
8ohm            -HepC Helicase

Length = 435


 Score = 46.1 bits (107), Expect = 7e-05
 Identities = 73/274 (26 Positives = 107/274 (38 Gaps = 19/274 (6


Query: 196 GAGKTKRYLPAIVREAIKRGLRTLILAPTRVVAAEMEEALRGLPIRYQTPAIRAVHTGRE 255
            G GK    A    K      A T    M A G      T  RA TG
Sbjct: 18  GSGKSTKVPAAYAAQGYKVLVLNPSVAATLGFGVYMSKA-HGIDPNIRT-GVRAITTGGP 75



Query: 256 IVDLMCHATFTMRLLSP-VRVPNYNLIIMDEAHFTDPASIAARGYISTRVEMGEAAG--- 311
            I      T    L      Y II DE H TD SI    I T    E AG
Sbjct: 76  IT----YSTYGKFLADGGCSGGAYDIIICDECHSTDSTSILG---IGTVLDQAETAGARL 128



Query: 312 -IFMTATPPGSRDPFPQSNAXXXXXXXXXXXXXSWNSGHEWVTDFKGKTVWFVPSIKAGND 370
             TATPPGS    P N                          G   F S K
Sbjct: 129 VVLATATPPGS-VTVPHPNIEEVALSNTGEIPFYGKAIPIEVIRGGRHLIFCHSKKKCDE 187



Query: 371 IAACLRKNGKKVIQLSRKTFDSEYVKTRTNDWDFVVTTDISEMGANFKAERVIDPRRCM- 429
              AA L  G     R  D     T    VV TD  G      VID  C
Sbjct: 188 LAAKLSALGLNAVAYYR-GLDVSVIPTSGD--VVVVATDALMTGYTGDFDSVIDCNTCVT 244



Query: 430 KPVILTDGEERVILAGPMPVTHSSAAQRRGRIGR 463
             V       I    P   S QRRGR GR
Sbjct: 245 QTVDFSLDPTFTIDTTTVPQDAVSRSQRRGRTGR 278
```

```
MyDengueProtein   AGVLWDVPSPPPMGKAELEDGAYRIKQKGILGYSQIGAGVYKEGTFHTMWHVTRGAVLMH  60
DENGUE            AGVLWDVPSPPPVGKAELEDGAYRIKQKGILGYSQIGAGVYKEGTFHTMWHVTRGAVLMH  60
                  ************:***********************************************


MyDengueProtein   KGKRIEPSWADVKKDLISYGGGWKLEGEWKEGEEVQVLALEPGKNPRAVQTKPGLFKTNA  120
DENGUE            KGKRIEPSWADVKKDLISYGGGWKLEGEWKEGEEVQVLALEPGKNPRAVQTKPGLFRTNT  120
                  ********************************************************:**:


MyDengueProtein   GTIGAVSLDFSPGTSGSPIIDKKGKVVGLYGNGVVTRSGAYVSAIAQTEKSIEDNPEIED  180
DENGUE            GTIGAVSLDFSPGTSGSPIVDKKGKVVGLYGNGVVTRSGAYVSAIAQTEKSIEDNPEIED  180
                  *******************:****************************************


MyDengueProtein   DIFRKRRLTIMDLHPGAGKTKRYLPAIVREAIKRGLRTLILAPTRVVAAEMEEALRGLPI  240
DENGUE            DIFRKRRLTIMDLHPGAGKTKRYLPAIVREAIKRGLRTLILAPTRVVAAEMEEALRGLPI  240
                  ************************************************************


MyDengueProtein   RYQTPAIRAVHTGREIVDLMCHATFTMRLLSPVRVPNYNLIIMDEAHFTDPASIAARGYI  300
DENGUE            RYQTPAIRAEHTGREIVDLMCHATFTMRLLSPIRVPNYNLIIMDEAHFTDPASIAARGYI  300
                  *********  *********************:***************************


MyDengueProtein   STRVEMGEAAGIFMTATPPGSRDPFPQSNAPIIDEEREIPERSWNSGHEWVTDFKGKTVW  360
DENGUE            STRVEMGEAAGIFMTATPPGSRDPFPQSNAPIMDEEREIPERSWNSGHEWVTDFKGKTVW  360
                  *******************************:****************************


MyDengueProtein   FVPSIKAGNDIAACLRKNGKKVIQLSRKTFDSEYVKTRTNDWDFVVTTDISEMGANFKAE  420
DENGUE            FVPSIKTGNDIAACLRKNGKRVIQLSRKTFDSEYVKTRTNDWDFVVTTDISEMGANFKAE  420
                  ******:*************:***************************************


MyDengueProtein   RVIDPRRCMKPVILTDGEERVILAGPMPVTHSSAAQRRGRIGRNPKNENDQYIYMGEPLE  480
DENGUE            RVIDPRRCMKPVILTDGEERVILAGPMPVTHSSAAQRRGRIGRNPRNENDQYIYMGEPLE  480
                  *********************************************:**************

MyDengueProtein   NDEDCAHWKEAKMLLDNINTPEGIIPSMFEPEREKVDAIDGEYRLRGEARKTFVDLMRRG  540
```

```
DENGUE            NDEDCAHWKEAKMLLDNINTPEGIIPSMFEPEREKVDAIDGEYRLRGEARKTFVDLMRRG 540

                  ************************************************************

MyDengueProtein   DLPVWLAYRVAAEGINYADRRWCFDGVKNNQILEENVEVEIWTKEGERKKLKPRWLDARI 600

DENGUE            DLPVWLAYKVAAEGINYADRRWCFDGTRNNQILEENVEVEIWTKEGERKKLKPRWLDARI 600

                  ********:*****************.:********************************


MyDengueProtein   YSDPLALKEFKEFAAGRK 618

DENGUE            YSDPLALKEF---AAGRK 615

                  **********.. *****
```

1df9A   -Protease

Length = 177

 Score = 322 bits (817), Expect = 4e-88

 Identities = 158/177 (89 Positives = 161/177 (90


```
Query: 5    WDVPSPPPMGKAELEDGAYRIKQKGILGYSQIGAGVYKEGTFHTMWHVTRGAVLMHKGKR 64
            WDVPSPPP GKAELEDGAYRIKQKGILGYSQIGAGVYKEGTFHTMWHVTRGAVLMHKGKR
Sbjct: 1    WDVPSPPPVGKAELEDGAYRIKQKGILGYSQIGAGVYKEGTFHTMWHVTRGAVLMHKGKR 60



Query: 65   IEPSWADVKKDLISYXXXXXXXXXXXXXXXXXVQVLALEPGKNPRAVQTKPGLFKTNAGTIG 124
            IEPSWADVKKDL S               VQVLALEPGKNPRAVQTKPGLFKTNAGTIG
Sbjct: 61   IEPSWADVKKDLVSCGGGWKLEGEWKEGEEVQVLALEPGKNPRAVQTKPGLFKTNAGTIG 120



Query: 125  AVSLDFSPGTSGSPIIDKKGKVVGLYGNGVVTRSGAYVSAIAQTEKSIEDNPEIEDD 181
            AVSLDFSPGTSGSPIIDKKGKVVG YGNGVVTRSGAYVSAIAQTEKSIEDNPEIEDD
Sbjct: 121  AVSLDFSPGTSGSPIIDKKGKVVGIYGNGVVTRSGAYVSAIAQTEKSIEDNPEIEDD 177
```

Molecular Biology Protocols.

BACTERIAL TRANSFORMATION:

1. 1 µl of plasmid DNA added to cells

2. Leave on ice for 45 minutes

3. Heat in waterbath at 42 °C for 2 minutes

4. Add 0,5 µl of LB medium with no antibiotic

5. Leave on ice for 1 hour at 37 °C

6. Microcentrifuge at 8000 rpm for 1minute

7. Take off and discard 0,5 ml of supernatant

8. Re-suspend gently in remaining volume

9. Plate on LB agar + AMP Petri dishes

E.COLI CULTURE

1. Pick up a healthy looking (round, consistent) colony with a loop

2. Inoculate in a universal tube with 10 ml of LB + AMP

3. Incubate in shacking centrifuge at 37 °C for 13 to 14 hours

MINI-PREP PREPARATION

1. Add 1,5 ml of overnight culture of E.Coli in an eppendorf and spin at 13000 for 1 minute

2. Repeat 3 times for each tube → passing a total of 4,5 ml of cell suspension from each tube

3. Each time discard supernatant

MINI – PREP

using a microcentrifuge

This protocol is designed for purification of up to 20 µg of high-copy plasmid DNA from1–5 ml overnight cultures of *E. coli* in LB (Luria-Bertani) medium.

Note: All protocol steps should be carried out at room temperature.

Procedure

1. Resuspend pelleted bacterial cells in 250 µl Buffer P1 and transfer to a microcentrifuge tube.

2. Ensure that RNase A has been added to Buffer P1. No cell clumps should be visible after resuspension of the pellet.

3. Add 250 µl Buffer P2 and gently invert the tube 4–6 times to mix.

4. Mix gently by inverting the tube. Do not vortex, as this will result in shearing of genomic DNA. If necessary, continue inverting the tube until the solution becomes viscous and slightly clear. Do not allow the lysis reaction to proceed for more than 5 min.

5. Add 350 µl Buffer N3 and invert the tube immediately but gently 4–6 times.

6. To avoid localized precipitation, mix the solution gently but thoroughly, immediately after addition of Buffer N3. The solution should become cloudy.

7. Centrifuge for 10 min at 13,000 rpm (~17,900 x g) in a table-top microcentrifuge.

8. A compact white pellet will form.

9. Apply the supernatants from step 4 to the QIAprep Spin Column by decanting or pipetting.

10. Centrifuge for 30–60 s. Discard the flow-through.

11. (Optional): Wash the QIAprep Spin Column by adding 0.5 ml Buffer PB and centrifuging for 30–60 s. Discard the flow-through.

12. This step is necessary to remove trace nuclease activity when using *endA+* strains such as the JM series, HB101 and its derivatives, or any wild-type strain, which have high levels of nuclease activity or high carbohydrate content. Host strains such as XL-1 Blue and DH5α™ do not require this additional wash step.

13. Wash QIAprep Spin Column by adding 0.75 ml Buffer PE and centrifuging for 30–60 s.

14. Discard the flow-through, and centrifuge for an additional 1 min to remove residual wash buffer.

15. IMPORTANT: Residual wash buffer will not be completely removed unless the flow-through is discarded before this additional centrifugation. Residual ethanol from Buffer PE may inhibit subsequent enzymatic reactions.

16. Place the QIAprep column in a clean 1.5 ml microcentrifuge tube. To elute DNA, add 50 µl Buffer EB (10 mM Tris·Cl, pH 8.5) or water to the center of each QIAprep

17. Spin Column, let stand for 1 min, and centrifuge for 1 min.

DNA CONCENTRATION PROTOCOL

Materials:

1. TE solution

   o   10 mM Tris (pH to 7.5)

   o   1 mM EDTA (pH to 8.0 to dissolve)

2. DNA sample

3. SYBR Green I(R) nucleic acid gel stain (Molecular Probes)

4. Plastic wrap

5. distilled water

6. DNA marker stock (10 mg/ml)

Supplies:

1. Tubes

2. Polaroid setup (with proper filter - SYBR Green/Gold gel stain photographic filter) and UV light box &

3. Micropipetter and tips

Procedures:

1. Prepare 6 DNA standards from DNA marker stock:

   o standard I (5 ug/ul): 1:2 dilution of DNA marker stock

   o standard II (2.5 ug/ul): 1:2 dilution of standard I

   o standard III (1.25 ug/ul): 1:2 dilution of standard II

   o standard IV (0.625 ug/ul): 1:2 dilution of standard III

   o standard V (0.313 ug/ul): 1:2 dilution of standard IV

   o standard VI (0.156 ug/ul): 1:2 dilution of standard V

2. Make a 1:5000 dilution of the SYBR Green I(R) with TE solution.

3. Mix 5 ul of the DNA sample and each of the 6 standards with 5 ul of the diluted SYBR Green I(R) dye.

4. Place a sheet of plastic wrap smoothly onto the UV light box.

5. Spot the mixtures individually onto plastic wrap.

6. Spot the set of 6 standards.

7. Turn on the UV light and take a photo (Polaroid 667 black-and-white print film).

8. Compare the brightness of the DNA sample with the DNA standards and estimate concentration.

LIGATION - PCR

- Add 1 µl of the vector

- Add 1 µl of 10x buffer

- Add PCR product (1 µl)

- Add 8 µl of SIGMA water

- Add 1 µl of Ligase

LIGATION – OVERNIGHT

- Add 0,5 µl of the vector

- Add 5 µl of 10x buffer

- Add 3,5 µl Insert DNA

- Add 1 µl of Ligase (last)

- Incubate overnight at 17 °C

LIGATION – BENCHTOP

- Add 0,5 µl of the vector

- Add 5 µl of 10x buffer

- Add 3,5 µl Insert DNA

- Add 1 µl of Ligase (last)

- Leave at room temperature for 2 hours + 30 minutes.

<u>SDS-Page Gel Preparation</u>

	10% Running Gel	Stacking Solution
Acrylamide	3.3 ml	696 µl
1.5M Tris-HCl pH8.8	2.5 ml	---
0.5M Tris HCl pH6.8	---	650 µl
10% SDS	100 µl	100 µl
10% APS	50 µl	50 µl
TEMED	20 µl	10 µl
Water (D)	4.0 ml	3.65 ml

→ The Running Buffer is made by preparing 200 ml SDS Page Buffer (containing 288g Glycine and 80g Tris). 20 ml of SDS 10% buffer into 2000ml of dH_2O will give the SDS buffer.

www.ingramcontent.com/pod-product-compliance
Lightning Source LLC
Chambersburg PA
CBHW031121180526
45160CB00005B/43/J